IPS

– a Forth-like Language for Space

In a later edition , the background with Forth and FORTH,INC. will be added or at least a link to the information will be provided.

Professor Dr. Karl Meinzer

Introduction to this book

A couple of weeks ago, I saw by chance a link to IPS. I had known about it for many years but did not know about more details.
I do like the language Forth – an unusual one I admit – but as a hardware person it fits better.

I contacted the people who had this information, including Prof. Dr. Karl Meinzer, and asked for permission to get the copyright from him just for a publication as eBook and as print book.

The answer was yes, and I got started.

Some decisions had to be taken about how to convert it, and the approach was to keep the original stucture in the book as much as possible, just reformat to fit into the book.
The original page numbers are still included, so people who own the original data can cross reference. The same applies to the index at the end.
While reading through it while converting, it became evident, how much work has gone into this IPS software project – probably one of the first Forth applications without using Forth – and used successfully many time in AMSAT Space Projects flying around our planet.
The book and feedback might unearth some additional information.
Especially, I would like to see a complete package and run the Startrek program not included here, but can hopefully be found now.
I have to thank Professor Dr. Karl Meinzer again for his permission.

Dipl.-Ing. Juergen Pintaske, Exmark, Exeter - May 2019 - IPS v5

IPS

HIGH LEVEL PROGRAMMING OF SMALL SYSTEMS

Second Edition

Karl Meinzer, Ph.D.
University of Marburg

1978

IPS - High Level Programming of Small Systems

First edition 1978
Second edition and first printing 1997.
Third edition (electronic PDF) with corrections 2016
Published by James Miller, Cambridge CB23 7PS, England.

Permission has been given by Professor Dr. Karl Meinzer and James Miller to Juergen Pintaske, Exmark to reformat and publish as eBook and as print book.

ExMark, May 2019

Contents iii

Page numbers on the left are like original, right as in this book

Contents

Contents v

Preface to 1st Edition

ARE YOU

owner of a microcomputer based on the 8080, 6800, 6502 or the CDP 1801/2 ?

ARE YOU

looking for an efficient, easy and clean way to program your computer using a high level language ?

DO YOU

wish to use your microcomputer in time critical applications like controlling devices, robotics or fancy games ?

DO YOU

wish to have your computer look after several tasks at the same time ?

THEN LOOK NO FURTHER !!!

IPS has been designed to solve these problems for the AMSAT space projects (communication satellites for radio amateurs) and has since been found to be a very useful tool for programming small systems.

It uses an extremely modular and structured approach to develop programs interactively. Because IPS is a high level language, it allows the sharing of programs regardless of the processor for which they were developed.

Preface to 2nd Edition

This manual was written by Karl Meinzer in 1978 during the evolution of the IPS operating system, but was never published in any substantive form.

A hand corrected draft printout was circulated to a few interested engineers who then reproduced it ad hoc, minus several chunks, via ever worsening photocopies.
But the paper original was declared lost.

Happily, in 1996 the IPS manual was discovered to have survived on Atari 800XL (ca. 1980) computer cassette tapes, and moreover had recently been transferred to floppy disc by Robin Gape, a prominent IPS contributor in the mid '80s.

Thus it became practical to republish the document.

In the late '70s, disc systems were uncommon. Thus the manual refers throughout to cassette tape as its input/output medium; IPS computers usually had
two recorders. This anachronism does not detract in any way from the relevance of IPS, and for the sake of consistency has been left in the document.

IPS development has not stood still. It has been written for the Atari ST and in 1996 for 32-bit Acorn RISC Computers. The latter are based on the ARM family of processors, and run IPS 100-1000x faster than the early 1802/6502/8080 systems. Hopefully re-publication of this manual will encourage versions for other
machines, notably the IBM-PC.

A significant application of IPS has been its use as the Operating System for AMSAT's Phase III series of communications spacecraft.

The Oscar-13 (Phase III C) satellite's 1802/IPS computer functioned without missing a beat for eight years until it re-entered the atmosphere in 1996. Impressive for any computer's OS, let alone one functioning in space.

This re-publication is dedicated to Dr Karl Meinzer DJ4ZC in recognition of 30 years outstanding contributions to the Amateur Space Programme.

James Miller, Cambridge, England, 1997.

Introduction

One of the most significant technological innovations during the last few years is the integrated microcomputer. This device is permeating all walks of life; it allows solution of a large number of engineering problems that previously could not be solved due to excessive hardware complexity.
With microcomputers the limiting factor has become the effort one is required to invest in software engineering.
There are two phases to each project.

1. The problem must be analyzed and the activities the computer is to perform must be defined.
2. A program has to be written and tested.

The work involved in point 1 can never be eliminated by programming aids, but the workload of point 2 is heavily dependent on the quality of the programming support used; with ideal support the work involved in point 2 should constitute only a small fraction of the total effort.

Most programming aids available fall far short of this goal. The conventional "high-level" programming aids are derived from the classical data processing environment with different goals and constraints than given with microcomputers, so it is not surprising that most microcomputer programming is still done at assembler level. This means that the problem must be broken down into the small steps the computer is able to execute, and one is thus forced to live with the sometimes quite unsystematic structure of a computer resulting from hardware constraints.

For some time it has been known that the usual mini- and microcomputer architectures (with their linear structure) are poorly matched to the "human" method of understanding and decomposing a problem. Humans tend to think in alternatives and hierarchies.
This eventually led to the programming concepts of "top-down design" and structured programming.

To date, there are no computers available matching these concepts. The next best thing one can do is to use a conventional microcomputer to simulate (emulate) such a high-level computer. The programming system IPS described in this book uses a general approach somewhat similar to FORTH (Moore [1]), but was designed to provide more demanding user interaction mechanisms and to use the typical low-cost peripherals of microcomputers.

IPS mostly is a high level language allowing extremely modular structured programming. In contrast to other languages it is essentially free of syntax rules. It uses RPN (reverse polish notation) to make parameter passing between modules extremely simple and is designed to be as unrestrictive as low-level assemblers.

The high-level emulation technique is extremely economical in terms of memory usage - the entire system resides in 6 Kbytes of memory.
But programs execute only two to three times slower than optimum assembler code. In most instances this represents no problem, however extremely time-critical applications or special hardware may require some assembly language interfaces. IPS thus employs an integral assembler to facilitate these extensions.

The design of IPS allows high level extensions as well. A detailed discussion of the philosophy of IPS is given at the beginning of chapter III.

This book is organized into three chapters of increasing sophistication.

Chapter I
presents an introduction to the language; the material presented there will enable you to accomplish all "regular" programming tasks.

Chapter II
introduces the assemblers for the COSMAC 1801/1802, the 8080, the 6800 and the 6502 allowing low-level extensions and interfaces.

Finally, Chapter III
Describes and documents the inner workings of IPS; this material will be useful if you wish to extend or change IPS itself.

The easiest way to learn IPS is to have access to a computer with IPS installed, so that you may "learn by doing" as you progress through this book.

Since IPS is an integrated programming system with no other software required, it necessarily needs a number of interfaces to the particular hardware. If your installation does not have the same configuration as the system for which IPS was designed, some adaptations may be necessary. These generally are rather trivial for a person familiar with IPS, but if this is your first exposure, it may be a good idea to procure the assistance of someone more familiar with IPS during the first installation on your system.

I am certain you will find IPS a useful tool for a vast number of programming projects.

If you should discover any errors or improvements, I will be pleased to learn of them.

Karl Meinzer
Marburg, 1978.

Acknowledgements

The development of IPS was triggered by the demanding and stimulating environment of AMSAT, a volunteer organization building communications satellites for radio amateurs as a public service with practically no funds.

Among the many individuals having contributed to IPS, I in particular wish to acknowledge the many helpful discussions and suggestions from
G. Groh who suffered with me through the first implementations.
J. Lipfert applied IPS to a number of real-time applications within the University of Marburg. His fighting with the many traps of early IPS resulted in
the present error philosophy.
R. Gerlich wrote a first implementation of IPS for the 6502. His work made many problems transparent particular to that processor, enabling the more compact version in this book.
R. Dunbar helped with the English mnemonics, the polishing of this book and (in 10 hectic nights) bringing up the 8080 IPS.

References

1. C.H. Moore Forth: A new way to program a
Minicomputer Astron. Astrophys.
Suppl. 15,497-511, 1974
2. T. Dollhoff Making hash with tables Byte, vol. 2,
No.1, 18-30, January 1977
3. J.F. Herbster Improving quadratic rehash Byte, vol.
2, No.5, 142, May 1977
4. W.W. Peterson Prüfbare und korrigierbare Codes
R.Oldenbourg Verlag, Munich and
Vienna 1967
5. N. Wirth Software Practice and Experience
Modula, (3 papers) vol. 7, 3-35,
37-65, 67-84, 1977
6. D.W. Johnson Multitask µP executive routine uses
only six instructions
Electronic Design 4, Feb. 15, 1978

CHAPTER I

How to use IPS

1 Hardware requirements for IPS

This book describes IPS for four microprocessors:

- The RCA COSMAC (CDP 1801/1802)
- The 8080 (thus also covering the Z80)
- The 6800
- The 6502

Hardware requirements

All four processors demonstrate similar performance in the IPS environment and require about the same amount of memory to support IPS. The system occupies about 6 Kbytes; in order to have sufficient space for user programs 16 Kbytes are recommended as the memory complement.
My experience indicates that the remaining space is sufficient to accommodate even very large programs. As a guideline, you may assume that IPS programs require about one half the memory of similar BASIC programs.

IPS is an extremely interactive system. Key to these interactions is a video memory organized as 16 lines by 64 characters. This memory is part of the processor address space; by storing into these addresses, the content is immediately displayed on a TV-screen.

(During processor access to this memory,
the TV-screen is blanked).

Some TV-memories on the market possess scrolling and selective blanking features to facilitate emulating a teletype mode. IPS does not use a teletype emulation mode and thus does not need nor use these features.

The TV-memory should at least be able to display the 64 ASCII character subset.

It should also allow inverting character and background brightness in order to be able to display a cursor.

If you utilize a hardcopy device having lower-case
capability, I recommend a display having the full 128 character set. This will allow you to edit normal correspondence or other written material in a generally acceptable form. inverse video.

To enter text into the system, an ASCII keyboard is required. The COSMAC and 8080 systems use a bit-serial approach for entering the characters, the other two versions enter the seven character-bits in parallel.

The 8080 may also be used with the parallel configuration. The parity-bit is not used.

The keyboard preferably should have the "CTRL" characters marked on the keys. Otherwise you will be required to mark some of the keys as to their control function.
Also, the keyboard should have a provision for locking it in upper case.

Make sure that the keyboard has the full ASCII character set; some keyboards do not have the @ character (which is frequently used in IPS).

Mass storage in the IPS versions described in this book is done using standard tape cassettes. The four systems use the AMSAT cassette standard.
(See appendix B).

From a software point of view the significant property of this standard is the fixed length of the records; all records are 512 bytes long and may contain arbitrary data (character or binary). This length allows the use of the TV-screen as an input buffer; one record (block) fills exactly one half of the TV-screen.

The AMSAT standard uses a synchronous rate of 400 bit/s and thus is approximately twice as fast as the asynchronous 300 bit/s Kansas-City standard.
The AMSAT standard is marginally acceptable from a speed point of view. If you do not intend to use the AMSAT recording technique, any other standard faster than this system and capable of blocks of 512 arbitrary bytes could be used;

of course the software drivers may have to be changed.

The IPS represents a multiprogramming system. It thus cannot use timing loops or other programming techniques which tie up the computer in a single task.

In order to have a means of coordinating the various activities and to keep time, a 20 ms hardware input of a sort is required. This input also allows the software clocks to keep time.
With the COSMAC this input is applied to " External Flag 1 "; with the other processors it forces an interrupt.

Finally, the system requires a bootstrap capability to enter IPS initially.

This may be a hardware load feature (COSMAC) or a small ROM program containing a loader.
Bootstrap loader

Note that no front-panel switches or indicators are required; all communication with the system is effected via the keyboard and the TV-display.

Only the cassette recorders need to be manipulated additionally.

Hardware requirements

Here is a summary of the hardware requirements for an IPS installation:

- **Processor :**
  CDP 1801, CDP 1802, 8080, Z80, 6800 or 6502

- **Memory :**
  16Kbytes RAM + loader ROM

- **Display :**
  1024 byte video-RAM (16 x 64)
  ASCII character set,
  bit 7 to invert background and character

- **Keyboard :**
  ASCII keyboard featuring full character set

- **Mass storage:**
  Cassette system enabling the recording of
  512 byte long blocks with arbitrary content
  e.g. the AMSAT cassette recorder standard

- **Timing:**
  20 ms crystal derived input to processor
  (ext. flag or interrupt)

- **Front-panel :**
  None required Hardware requirements

Chapter III contains information on how to install IPS on computers using other microprocessors than the ones mentioned. You may assume that in a system like IPS most good points and bad points of the various processors average out, so the performance variations between the different 8-bit-processors are not very significant. A better guideline for the choice of a system would be hardware availability and simplicity (chip-count).

Nevertheless, a processor should meet certain minimal criteria. Most important, it should have 16-bit addressing capability (at least 15-bit) without any data addressing discontinuities such as paging.

Having to handle such discontinuities would considerably slow down the system. Preferably the processor should incorporate some 16-bit registers on the chip allowing indirect addressing.
Ideally they should have an auto increment feature and indexing capability.

The COSMAC architecture comes closest to these requirements.
The 8080 is less well matched to IPS, but the powerful 16-bit instructions are a compensation.

With the 6800 operations requiring multiple 16-bit pointers are awkward, but the capability of indexing with reference to the stack is very powerful.

The 6502 suffers from not having a single 16-bit register on the chip. The high speed of this processor and the additional addressing modes nevertheless make this processor about as fast as the other ones. But the emulator code is significantly more voluminous than that of the other processors.

IPS also could be installed on 16-bit computers. It will run there about 2 to 5 times faster than on 8-bit machines. I doubt that there are many installations where this speed would actually be required. On the other hand, cost goes up nearly by a factor of two; presently memories are only available as 16K by 1, but not as 8K by 2.

Also the choice of the processor becomes more critical; processors well matched to a stack-organization (e.g. PDP 11) may perform significantly
faster than processors designed for a workspace concept.

2 Keyboard Operations

2.1 Your computer and IPS

Your computer, by its design, requires the instructions it is to perform as a binary pattern in its memory. These instructions are difficult for humans to handle, not only because of the binary nature, but also because of the linear arrangement in the computer memory. A belated insertion, for example, requires that the rest of the memory contents be modified.

Programming is much simplified, if we can instruct the computer in a language which matches our thinking. In addition, programs thus become self documenting, which eliminates the need for separate documentation (which is a pain in the neck and thus is often badly neglected).

We also need a simple and transparent means of testing our programs. IPS is designed to provide these services and to translate your program into a format which the computer can execute.

IPS uses a kind of dictionary. This dictionary contains the name of an activity and a description of the steps constituting that activity. IPS itself is largely such a dictionary, defining the translation process.
Each process the computer can execute may be described by such a dictionary. Creating such a dictionary describes to the computer how to handle the process.

The computer can interpret this dictionary, hence the name Interpreter for Process Structures (IPS). Most of this book describes how to create such a dictionary by merely explaining to the computer the problem you wish to solve.

2.1.1 Communicating with IPS

The following sections assume that you have loaded IPS into your computer.

The TV-screen has 16 lines with 64 characters each. In the lower 8 lines the computer expects its instructions from you via the keyboard or from cassettes.

The cursor (a black bar) indicates the next position you may write into. The eight upper lines are used by the computer to communicate back to you. Separator

All input to the computer consists of words; they are separated by one or more blanks. Words may consist of arbitrary strings of characters without blanks.

There are three different kinds of words:
numbers,
IPS-words and
words you have defined yourself.

After you have written some words, and if you have made no typing error, you press CR (Carriage Return) to tell the computer to process the words.

If the computer can process the words without error, it clears your entry to indicate that it has complied with your text.

If you made a typing error, you may correct it before pressing CR by using the following keys:

BS (CTRL-H) Backspace, moves the cursor 1 position back

FS (CTRL-\ ) Forwardspace, moves the cursor to next position

LF (Line Feed) Moves the cursor 1 line down

You cannot move the cursor off the screen; if you try to move it beyond the last line, it will show up in the first line and vice versa.

You correct your writing by simply overwriting the erroneous parts. The division of the screen into lines of 64 characters is invisible to the computer; you may write words over line boundaries without separation signs.

The computer "sees" the screen as a single line of 1024 characters.

After the CR, the computer reads the text from the beginning of the 8-th line to the CR.

You may write into the first 8 lines as well, but the computer ignores this text.

2.2 The Stack

Before you attempt to make the computer perform simple operations, the stack should be introduced. It is a temporary storage for numbers. You may visualize it as a number of cards, each having one number written on it; these cards are then "stacked" on top of each other in the sequence they are written. The last number thus is on top.

Most instructions of IPS expect some numbers on the stack; they remove these and replace them by results. Instructions expecting one number always refer
to the top of the stack, instructions expecting two numbers refer to the two top numbers and so on.
The results are then placed on top of the stack. Stack top of te stack

You will see that such a stack is an ideal place to store and to pass intermediate results within program modules as well as between them.

To simplify program development and debugging, the contents of the stack are displayed in the
first two lines of the TV-screen. The stack contents are displayed with the top entries to the right. If there are more than 16 stack entries, only the 16 topmost are shown. The stack display may be switched off. Parameter stack

2.2.1 Numbers

All numbers used by IPS are stored internally as 16-bit integers and thus range from -32768 to 32767.

Other number types may be defined if necessary.

The internal representation of the numbers is binary, with positive numbers bit 15 is zero;
with negative numbers bit 15 is set to one.

Logically they are produced by counting backwards from zero (-1 thus corresponds to

`1111111111111111` , or in hexadecimal writing **#FFFF**) binary number

2.2.2 Number entries

In order to place a number on the stack, you simply type it on the keyboard.

a. Unsigned numbers are read by IPS as positive decimal numbers and placed on the stack. (a + sign is not allowed)

b. A - sign followed by figures is read as a negative decimal number.

For example:

`123 -415 <CR>` produces a stack display

of `123 -415` `<CR>` will now be tacitly assumed to follow all entries.

You may remove numbers from the stack one at a time by typing the word DEL . DEL DEL will thus remove both of the above numbers.

c. In addition to decimal numbers you may also enter binary or hexadecimal numbers. Hexadecimal numbers are numbers starting with the # and having the digits 0 - 9 and A - F (10 - 15).

Binary numbers start with a B followed by digits of 0 or 1 .

With all numerical inputs, leading zeros may be omitted. But you must be careful not to exceed the permissible range possible with 16 bit numbers.

Otherwise, erroneous numbers are placed on the stack. The largest permissible numbers are:

```
      -32768           to       32767
       #0000           to       #FFFF
B0000000000000000 to B1111111111111111
```
(16 digits after B)

The format of the stack display on the TV-screen depends on the number last entered. For example:

15 B111 #15 results in the display:

`#000F #0007 #0015` (the last entry was in hexadecimal)

If you enter now -40 you get:

`15 7 21 -40` (the last entry was in decimal)

If the last entry was binary, the stack is displayed in hexadecimal; there is no binary display.

2.2.3 Arithmetic operations

The basic arithmetic operations are effected by the words:

`+ - * / /MOD`

If you use the above stack content

`( 15, 7, 21, -40 )`

and enter the "word" +

you obtain: `15 7 -19`

The upper two numbers of the stack were removed, added and the result was placed on the stack.

Entering a - now results in: **15 26**
Now entering a * results in: **390**

All arithmetic operations are modulo 216.

There is no diagnostic at overflow;
it is the programmer's responsibility to ensure that the number range is not exceeded.

The division and /MOD are defined for positive numbers only, but these may extend up to **#FFFF**.
/MOD is a division leaving a quotient and on top of it the remainder.

2.2.4 Logical operators

In addition to the arithmetic operations IPS enables logical operations; there are three operators (words):

AND OR XOR

These operations expect two numbers on the stack just like the arithmetic operators; the operation takes place bit by bit and the result is left on the stack.

With **AND** the bits have to be pairwise 1 in order to have the corresponding bit in the result to be 1.

With **OR** at least one of the bits has to be 1 to have a 1 in corresponding bit in the result.

With **XOR** a 1 in the result is obtained, if the two corresponding bits are different.
e.g.: **B1010 B1100 XOR** results in **#0006** (i.e. **B0110**)

So far we have seen only operations using two numbers. There are also operations using a single number.

The word **INV** replaces all bits of a number by the opposite.

The word **BIT** takes a number from the stack and discards all bits except the four least significant ones. This resulting number (it may range from 0 to 15 or #0 to #F) is taken as a number identifying a specific single bit in a 16-bit number.

(The 16 bits of a number may be identified as bits 0 through 15, bit 0 being the least significant.) **BIT** now places a number on the stack in which all bits are 0 except the one identified by the original four least significant bits.

e.g.: **#7 BIT** results in **#0080** equal **B0000000010000000**
Bit-number: 15 7 0

This function is useful, if you want to test large numbers of bits in loops.

2.3 Stack manipulations

2.3.1 Parameter Stack Operations

So far you can see that the stack is the source and destination of all operations involving numbers or logical quantities.
In order to give you more freedom in dealing with these quantities, IPS provides some stack manipulation
operators. You have already learned one; the word DEL .

This is the complete list:
Stack manipulation
Parameter stack
Stack operators
Operators stack

DEL	DELetes the top entry
PDEL	DELetes the two top entries (a Pair)
DUP	DUPlicates the top entry
PDUP	DUPlicates the two top entries (a Pair)
SWAP	interchanges the two top-entries
RTU	Rotates the Three top entry Up, the old topmost entry becomes the third-lowest
RTD	Rotates the Three top entries Down, the old third-lowest becomes the top entry
SOT	duplicates the Second entry On Top of the stack.

The following table demonstrates these operations:

	Before operation	**After operation**
SWAP	5 8	8 5
DUP	8 5	8 5 5
DEL	8 5 5	8 5
RTU	8 5 12	12 8 5

	Before operation	**After operation**
RTD	12 8 5	8 5 12
SOT	8 5 12	8 5 12 5
PDUP	8 5 12 5	8 5 12 5 12 5
PDEL	8 5 12 5 12 5	8 5 12 5

2.3.2 Return Stack Operations

In addition to the normal stack (the operand or parameter stack) there is another stack IPS uses internally. This stack is called the return stack for reasons explained later; the return stack may also be used by the programmer.

There are some restrictions to be observed though, that will be explained in the section describing loops, (Sect. 3.1.2).

The return stack often represents a welcome intermediate storage area, if you wish to access deeper layers of the stack.

These three operations are available:

S>R takes a number off the stack
and places it on the return stack

R>S takes a number off the return stack
and places it on the stack

I (called Index, see Sect. 3.1.2)
duplicates the content of the return
stack onto the stack without changing the return stack.

This table demonstrates the actions:

		Before operation	**After** operation
S>R	S:	5 7	5
	R:	empty	7
I	S:	5	5 7
	R:	7	7
R>S	S:	5	5 7
	R:	7	empty

Note: Do not remove numbers from the return stack that you did not place there previously !

2.4 Named constants and variables

The stack is very practical for short term storage and for the manipulation of numbers, but it is not very suitable for the permanent storage of numbers.

IPS offers two possibilities to store numbers under a name: constants and variables.

To accomplish this, IPS creates an entry in the dictionary consisting of a compressed form of the name (namecode, 4 bytes), a pointer to code (2 bytes), an identifier of the code type (typecode, 2 bytes) and the value of the entry (2 bytes).

This structure is fully described in Chapter III.
See also Sect. 2.5.

Constant entries are produced for those numbers which are not changed during the course of the program.

To create a constant entry, you enter the value the constant is to have, the word **CON** and the name the constant is to have.

e.g.: **510 CON LIMIT** produces a constant-entry named LIMIT containing a value of 510.

If you now type the word **LIMIT**, the constant **510** is placed on the stack.
(It also continues to exist in LIMIT, of course.)

You may use any string of arbitrary characters for the names of dictionary entries - that includes special and control characters.

For example **$** or **>>>** are perfectly valid names.

Only blanks and round brackets are not permissible.
(See Sect. 3.1)

Because changing the value of constants is a bit involved, there is a second type of number-entry called variable.

e.g.: **745 VAR AUTO** creates a variable entry.

Typing AUTO does not deliver the value of AUTO, but the memory position where AUTO is stored. Variables thus are not named numbers, but namedlocations (addresses) in which numbers may be stored.

In order to obtain the contents of such a location the word @ is used. It replaces an address on the stack by the actual data residing at that address.

e.g.: **AUTO @** results in **745** to be placed on the stack

(Please remember that words are separated by one or more blanks.)

Because the variables deliver the address on the stack, it is simple to change their value.
The word ! (store) expects two numbers on the stack; a number that is to be stored and on top of it the address where it is to be stored.
When performing storage, the ! removes both numbers.

e.g.: **LIMIT AUTO !** calls the constant LIMIT (510), calls the address of AUTO ; then stores the 510 into AUTO. Now entering AUTO @ delivers 510 to the stack.

It is also possible to change the value of a constant; it is recommended that this option be used during program development only, however.
(Some IPS versions store constants in ROM).

In order to store a new value into a constant, you need the address of that constant. The word **?** followed by the name of the constant delivers this address to the stack.

e.g.: **45 ? LIMIT !** puts a 45 on the stack, then the address of LIMIT is placed on the stack and finally ! stores the 45 into the address of LIMIT .

Some IPS versions do not allow use of the **?** in programs. It is good practice to create variable entries if the value of that entry changes during the program.

2.5 Fields

Often a set of numbers set required, which are stored under a single name and which differs only by an index.
Such structures are called fields. Such fields may be defined by the use of an extension of the variable concept. In order to understand this, let us take a closer look at how variables are stored in memory.

All entries in the dictionary (nametable) are very similar.

The nametable contains the "name" coded into 4 bytes. The first byte contains the number of characters in the name (truncated to 63, if necessary).
The following three bytes contain a 24-bit number computed from all the ASCII

characters of the name. All entries into the IPS dictionary utilize this name coding; each name has a unique code and may thus be identified later.

In the nametable each namecode is followed by a pointer which indicates the dictionary position containing the typecode of the entry.

A variable entry in the dictionary consists of a 2 bytes typecode for variable followed by 2 bytes containing the actual numerical value of this particular variable.
For full details see Chapter III, Sect. 5.3.

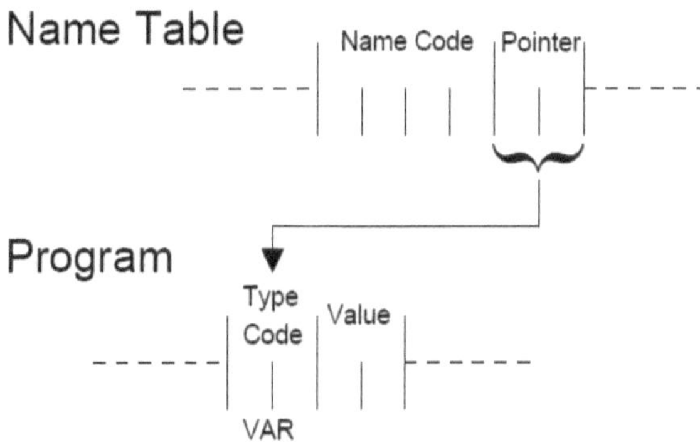

IPS has a system-variable named $H, which contains a pointer to the next free position in the dictionary. Thus, IPS keeps track of where the next entry in the dictionary is to go.

If you increment this pointer after creating a variable-entry, you make room in the variable to contain more than two bytes.
If you want to create a field for 10 numbers, for example, you must increase $H by 9 number positions (18 bytes total, since you will recall that each number entry consists of two bytes.)

e.g.: `0 VAR GROUP` produces a variable GROUP

If you want to be able to store into it 10 numbers, you must increase `$H` by 18 (bytes).

```
$H @ 18 + $H !
```

(instead of writing $H @ you may also write **HERE** ; it has the same effect)

To now store a value of 15 into component number 4 of the field (counting: 0 1 2 3 ...), you type:

```
15 GROUP 8 + !
```

The 8 + increases the address by 8 bytes, that is by four numbers, before the storage takes place.

Similarly, component 3 is accessed by:

```
GROUP 6 + @
```

2.5.1 Field Operations

To simplify the creation and manipulation of fields, IPS offers five words;

FIELD !FC >>> L>>> and F-COMP

To create a field you may write:

8 FIELD ALPHA

This creates a field named ALPHA being 8 bytes wide.

In this field all numbers are initially undefined.
To preset them to specific values, you type:

5 9 -123 17 ALPHA 4 !FC (store field components)

This operation stores the 4 numbers 5 9 -123 17 (in this order) into the field ALPHA ; the number of components must always be specified.

With the operation **!FC**, as with all other IPS operations expecting numbers on the stack, the following principle is always observed: first the "source", then the "destination" and finally additional information is expected on the stack in this order.

To move components of a field into another field, you could accomplish this component by component using @ and ! .

With a large set of data this would be clumsy and slow. To simplify and speed up this activity, a special word is available.

Using the two fields **GROUP** and **ALPHA** of 20 and 8 bytes, respectively, you might move ALPHA into GROUP say, beginning at position 10:

ALPHA GROUP 10 + 8 >>>

source destination no of bytes field transport

Graphically:

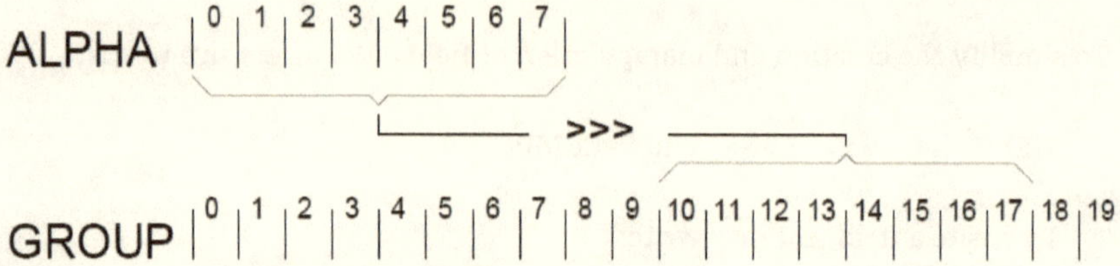

ALPHA | 0 | 1 | 2 | 3 | 4 | 5 | 6 | 7 |

\>>>

GROUP | 0 | 1 | 2 | 3 | 4 | 5 | 6 | 7 | 8 | 9 | 10 | 11 | 12 | 13 | 14 | 15 | 16 | 17 | 18 | 19 |

This transport takes place directly (not via the stack) and thus is much faster than transporting individual components.

The transport takes place byte by byte starting at the lower addresses.

The addresses may overlap. This enables the following useful function:

Assume you wish to set a total field to zero, you type:

```
0 ALPHA !        ( clears the first two bytes )
ALPHA ALPHA 1 +                          7 >>>
```

Graphically:

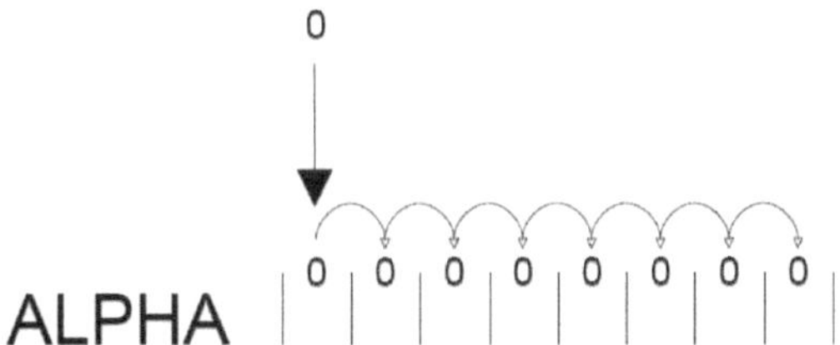

The word >>> transports 1 - 256 bytes.
If you wish to transport a larger number of bytes, you may use the word L>>> .
This word transports between 0 and 32767 bytes.
If the number of bytes to be transported is negative, no transport takes place.

The word L>>> takes slightly longer to execute than >>> . For short transports the word >>> thus is the preferred one.

In this context note that there is also an instruction which compares two fields, byte by byte (length: 1 - 256 bytes).

Write:

GROUP	**ALPHA**	**8**	**F-COMP**
field 1	field 2	no. of bytes	field comparison

This operation compares the first 8 bytes of the two fields.
If these field contents are identical, F-COMP leaves a 1 on the stack.
If GROUP is numerically smaller than ALPHA , the result is 0;
if GROUP is larger, the result is 2 .
The bytes having the higher addresses are also treated as the more significant ones.

2.6 Byte manipulations

Numbers are stored in two bytes, and the numerical or logical operations always operate on 16-bit (2-byte) quantities. Often it is desirable though, to access single bytes (8 bits) particularly if you wish to manipulate text strings.

There are two words enabling byte accesses:

@B expects an address on the stack; it removes it and replaces it with the content of the byte at that address. This number may have values between 0 and
255 (#00 to #FF) and is the least significant byte on the stack. The most significant byte on that stack entry is set to 0 (#00).

!B stores a single byte into an address.
The operation is analogous to the ! ;
the most significant byte of the number to be stored is discarded.

In variables and constants, the byte having the lower address is also the least significant one.
Numbers on the stack always are held and handled as two bytes - there is no single byte access to the stack.

2.7 Error messages

If IPS does not recognize a word or a number, an error indication is produced. IPS at first checks whether the input word is already in the dictionary.

If so, the word is executed.
Else IPS checks whether the input word is a valid number.
In this case the number is placed on the stack.

If both of these tests fail, IPS writes in front of the unidentified word, an inverted ? and stops processing the input.

You may now correct (starting after the ?) your text.
If you enter CR again, IPS continues to read after the ? .

There are a few activities, in which the instruction is followed by a name (e.g. VAR). If there is an error in such an operation, both words must be repeated.

Often there is a longer text following the error ? . Make sure that you place the cursor after the last word of your entry before pressing CR.
IPS reads only to the position of the cursor prior to entering CR .

There are only a few error messages beyond the ? mentioned; they are written into the seventh line:

STACK EMPTY !	You removed a number from the empty stack; the stack had to be restored.
MEMORY FULL !	The dictionary cannot accept further entries. This will rarely be encountered.

DUPLICATE NAME !	The name is already in use by you or the system; select a different one.
TEXT-ERROR !	)
STRUCTURE ERROR!	) These will be explained in Section 3.
NAME MISSING !	)

The error messages remain on the screen until they are acknowledged by typing OK.

This will clear the message and also re-enable cassette reading.
(An error will stop it, see Sect. 3.3, Tape Operations)

3 Programming

3.1 Definitions

So far, methods have been introduced which allow the computer to perform mathematical or logical operations.
The main strength of a computer though, lies in its ability to "remember" a sequence of actions and to perform them at a later time as dictated by other events, e.g. at a certain time or after certain external events.

To enable the computer to remember a set of actions, select a name and then enter the actions required. To be precise, type:

```
: name word word ... word ;
```

The colon produces an entry identified by the name which follows the colon.

The following words are not executed, but stored in the entry for later execution (coded internally as addresses). The semicolon terminates the entry.

Again:
words are separated by blanks; the semicolon is a word, too.

In order to make such definitions of actions more readable, you may insert arbitrary comments - simply enclose these comments by parentheses. IPS will ignore these comments, but later they may help you to understand your programs.

Example of a definition:

```
: 4NUMBERS 10 DUP ( PUTS 10 ON THE STACK TWICE )
           15 DUP ( PUTS 15 ON THE STACK TWICE );
```

If you now type 4NUMBERS , the stack will display:

```
10 10 15 15
```

You may see something new from this example beyond the creation of definitions: within definitions numbers are also not executed (put on the stack), but "compiled" into the definition in such a way that they will be put on the stack at the time of actual execution of the definition.

The only rule which must be observed is that you can put into definitions only items as are defined within IPS itself, or those which you have defined previously.

Of course, you may put into definitions other definitions you created before; thus you may create arbitrary hierarchies of definitions.

3.1.1 Branches

One of the most important properties of computers is their ability to make decisions based on the result of calculations or other events. IPS offers several possibilities towards this end. They assume that you have left a number on the stack. This number is to govern the further actions of the computer. For the time being, let us assume that there is a 0 or a 1 on the stack. You may then use the following words:

YES?	**word word ... word**	(actions to be performed, if a 1 was on the stack)
NO:	**word word ... word**	(actions to be performed, if a 0 was on the stack)
THEN	**word word ...**	(continuing action in both cases)

The word YES? removes a number from the stack and checks it.
If it is a 1, the words following the YES? are executed and then the words following the THEN.
If there was a 0 on the stack, the words following the NO: are executed and the program continues after the THEN.

Rule: Each YES? must be followed by a corresponding THEN
The NO: may be omitted; if the number on the stack is 0, the program continues after the THEN.

Generally there is not a 0 or 1 on the stack to base a decision on, it rather has to be calculated or produced by a comparison.

Strictly, only the least significant bit of the number is checked. An odd number leads to the YES? action, an even number results in the NO: action.

IPS offers the following standard comparisons:

```
>0    ) These compare a number on the stack against 0. They
<0    ) remove the number and replace it by a 1, if the
=0    ) condition is true, else they put a 0 on the stack.

>     )
<     ) These compare two numbers on the stack and remove them.
=     ) They put a 1 on the stack, if the condition is
>=    ) true, else they put a 0 on the stack.
<=    ) operators comparison comparison operators
=/    )
<>    ) Equivalent to =/ in some versions of IPS.
```

F-COMP Compares two fields, as explained earlier.

In addition to these standard comparisons, you may also use the logical operations. In this case you utilize only their effect on the least significant bit.

e.g.:
```
A @ B @ > C @ D @ < AND
            YES? ...
            NO:  ...
            THEN ...
```

These words will test, if A > B and C < D . Only if both are true, the words following YES? are executed.

It is a good exercise to write down the state of the stack after each word.

The "aligned" method of writing the YES? NO: THEN of course is not mandatory, but it helps to improve the readability of programs and is an insurance against trivial errors (like forgetting the THEN).

3.1.2 Loops

Many problems require a repetition of actions before you may continue with other instructions. IPS has four ways of effecting repetitions.

1. Instruction sequences requiring an initially unknown number of repetitions are written between the words LBEGIN and LEND? . LEND? expects a number on the stack and removes it.
If it was 1, the loop terminates and the computer proceeds with the activities following the LEND? .
If the number found by LEND? was 0, the instructions following the LBEGIN are repeated again. This repetition continues until the LEND? finally finds a 1. (Again, as with the YES? , only the least significant bit is tested; the loop terminates with an odd number.)

The following program demonstrates this concept.

```
: PICTURE TV4 ( THIS IS THE ADDRESS OF THE 4TH LINE )
        LBEGIN #40 ( ASCII FOR @ )
                SOT ( GETS ADDRESS, INITIALLY TV4 )
                !B ( DEPOSITS @ INTO THE ADDRESS  )
                1 + ( INCREMENTS ADDRESS )
                DUP TV4 64 + = ( ADR. = TV4 + 64 ? )
        LEND?   DEL ( ADDRESS ) ;
```

This program may now be executed by typing PICTURE . It will fill line 4 of the TV with @-characters.

Note: IPS has the constants TV0 , TV4 and TV8 containing the addresses of the start of line 0, 4 , 8 of the TV-screen.

The constant $TVE is the address of the last character on the TV.

2. With the LBEGIN ... LEND? construct the loop is executed at least once, because the termination test is at the end.
If you need a loop with the test at the beginning, use this construct:

```
LBEGIN      words for the test
YES?        words to be executed, if test delivered 1
THEN/REPEAT
```

The THEN/REPEAT always transfers control back to LBEGIN. The loop terminates if the test prior to YES? delivers a 0.

Again, a small sample program:

```
: A-LINE TV4 LBEGIN
             DUP TV4 64 + < ( LINE NOT YET COMPLETE )
           YES?
             #41 ( ASCII FOR A ) SOT !B   1 +
           THEN/REPEAT   DEL ;
```

This program will fill line 4 with As. operators looping looping operators

In the LBEGIN ... YES? ... THEN/REPEAT construct the word NO: is not permissible. Logically the same result will be obtained, if the NO: action follows the THEN/REPEAT.

3. Actions to be performed a specific number of times, are placed between the words EACH ... NOW . This construct has an iteration counter (also calledindex).
The word EACH expects two numbers on the stack, the loop begin (the initial value of the iteration counter) and the loop limit. The loop is executed for each consecutive number of the iteration counter until it exceeds the loop limit.
You may simplify the sample program of above by writing:

```
: CLEAR TV4 TV4 63 + EACH #20 ( ASCII BLANK )
                          I ( GETS THE INDEX, IN THIS
                              CASE THE ADDRESS )
                          !B
                          NOW ;
```

CLEAR will clear line 4 .

The EACH takes the two parameters off the stack and puts them on the return stack.

The NOW increments the iteration counter and then checks whether

it exceeds the loop limit.
If not, the loop (between EACH and NOW) repeats.
Else, the NOW discards the two parameters on the return stack and the loop is terminated.
Inside the loop the loop limit and the iteration counter are on the return stack; the later on top.

You may thus use the word I to access the iteration counter. If the initial value of the iteration counter exceeds the limit, the loop action is not executed at all. loop index operators looping looping operators

4. If you desire another increment than one with the loop index, you may use the word +NOW instead of NOW. +NOW expects a number on the stack and removes it.
The iteration counter is incremented by this number. e.g.

```
: PATTERN TV4 TV4 63 + EACH #5F ( ASCII FOR _ )
          I !B
          3 +NOW ;
```

will write an "_" into each third position when you type PATTERN .

The return stack first belongs to IPS; if you wish to use it, you must comply with these restrictions:

a. If you place something on the return stack within a definition you must remove it again in the same definition. IPS uses the return stack to remember where it is to return after executing a definition.

b. The EACH ... NOW construct also uses the return stack. If you place a number on the return stack within the EACH ... NOW loop, you also must remove it inside the loop.

If you call a definition from within the loop, note that this definition puts a number on the return stack. Thus you cannot call I within this definition. If you need I within the definition, put it on the stack prior to calling the definition.
(These rules may be circumvented by appropriate stack manipulations. But programs become difficult to read this way and this practice is therefore not recommended.)

3.1.3 Nesting rules

Constructs using the YES? NO: THEN and loops may be arbitrarily nested.
If you nest EACH ... NOW loops note that only the loop index of the inner loop is available by I. If you need the loop index of an outer loop, you have three options;
you may "unpack" the return stack,
you may put the index in the outer loop on the stack or
you may store the index in the outer loop into a variable.
Which of the solutions is optimum depends on the particular problem. Usually the last one is the most straightforward solution.

Make sure that each YES? has a corresponding THEN, each EACH has a corresponding NOW or +NOW and
each LBEGIN has a corresponding LEND? or THEN/REPEAT.

If you violate this rule, IPS will write
STRUCTURE ERROR! and the definition is not deposited into the dictionary.

STRUCTURE ERROR! with a residual number on the stack indicates a missing THEN, NOW or LEND? .
If you also get the message STACK EMPTY ! , there is an excessive THEN, NOW or LEND? .

It is good practice to define definition entries with an empty stack, only in this way are you guaranteed that a structuring error will not result in additional damage to your program. After a structuring error therefore, remove residual numbers from the stack with DEL .

Entries expecting a name (e.g. VAR) must receive the name during the same input. Otherwise you will get the message NAME MISSING and the action is not executed.
Apart from this rule, definition entries may extend over multiple inputs. In this case there will appear numbers on the stack, which you may ignore. The first number is always the same on your system and you may consider it an indicator for the definition mode of IPS. definition very long

3.2 Text handling

Besides handling numerical or logical quantities, IPS also must be able to communicate with humans. Therefore, there are a few words to allow the handling of text, particularly for output to the TV-screen. Basically text is treated similarly to numbers, that is; you may include your text output in definitions or you may put it into fields.
Each character occupies one byte. There are no basic differences between variables, fields containing numbers, or fields containing text strings except for the length of the parameter field.
With variables it is two bytes,
with fields you must define the length. text handling

Numbers you put into a definition are placed on the stack when you execute the definition. Text strings, on the other hand, will be written on the TV-screen.
There is a variable SP (screen pointer) whose content points to the position on the TV, where the next writing is to go. After writing, SP points to the first position following your writing.

Text strings are marked as such by quotation marks, e.g.

```
: TEST " THIS IS A TEST " ;
```

will create a definition, whose sole purpose is to write THIS IS A TEST . If you type now TEST, IPS will write into line 4 THIS IS A TEST . (SP is initialized to TV4.)

Entering TEST once more will append this text to the old one, and so on.

Since the TV-screen must be reused, there is a clear-word: BLANKS . This word expects a number on the stack and writes this number of blanks beginning at SP. SP is not changed.

You may set SP again to line 4 and clear the screen by typing:

```
TV4 SP ! 256 BLANKS
```

If now you call TEST, the game starts all over again.

If a text string is used by different program parts, it is more economical memory-wise to place it into a text array. If you are not in the definition-mode, a text string as input (in quotation marks) will leave on the stack the position of the first character on the screen and the number of characters in the string. (Blanks are characters too, of course). You may then store a text string into a field using !T e.g.:

```
" MOTOR DEFECT " GROUP !T
```

implies that there is a field named GROUP having a length of at least 12 bytes. MOTOR DEFECT will be stored into these 12 bytes, the rest remains unchanged.
If GROUP is 20 bytes long, and you want to fill the rest by blanks, you may write:

```
" MOTOR DEFECT                            " GROUP !T
```

Caution: If the text contains more characters than the field has bytes, IPS will be destroyed. The number of characters of a text-string is the number of positions between the quotation marks minus two.

You may create the field in conjunction with the text input.

```
" COLD "          4 FIELD CD          CD !T
```

Instead of writing 4 you might also have written DUP, but the length of the array would not be documented that way.

Text strings may have a length between 1 and 256 characters. If you forget the closing " or if the string is too long, you will get the diagnostic message TEXT-ERROR ! and the stack indicates the length of the erroneous string. Within a definition this also will result in a structuring error.

Outside definitions, the text string and the words connected with the further processing of the text must be within the same input, since the screen will be cleared after the input and thus the text will not be available beyond the input.

Text stored in fields is transported to the screen using the word WRITE .
WRITE expects on the stack the address of the field containing the string and on top the number of characters to be written.

Again SP is used as pointer, and after writing, SP points to the next free position.

```
CD 4 WRITE
```

will write COLD on the screen. (See previous example).

To write numbers on the screen, you may use the word CONV . It takes a number from the stack, converts it into an ASCII string and then writes it at SP .
SP again is updated to the next free position. The format of this writing is identical to the stack display.

The presentation (decimal or hexadecimal) depends on the last input. If you want to make certain, that the display is either decimal or hexadecimal, you must store (prior to calling CONV) a 10 or a 16 into the variable BASE .

e.g.: `10 BASE !`

before calling CONV will guarantee a decimal display.
CONV may produce up to six characters.

Text strings of course also may be transported to the screen using >>> . In this case SP , if you use it, is not updated.

Also you may produce text on the screen by explicitly storing the numerical representation of the ASCII characters as was demonstrated in the loop sample-programs.

All techniques together give you the choice to create the display handling best suited for your application.

3.3 Tape cassette operations

Programs (definitions) so far were only entered manually. But you may store your typing on cassettes. The recording technique uses low-cost cassette recorders and tape cassettes. The technical details are described in appendix B.
Each recording operation records a "block" of 512 bytes at a time (exactly one half TV-screen, 8 lines). The recorder should be modified in such a way that the computer can start and stop the recorder motor.

This feature allows automation of the more important tape operations.

To create a text recording, you first put IPS into the text editing mode by entering the word TEXT . Then you have a number of additional editing keys at your disposal.

1. CTRL P (DLE) clears the rest of the screen after the cursor. The cursor then is positioned at the beginning of the screen. Hitting DLE twice thus will clear the whole screen.

2. CR (Return) will not turn the text over to IPS, rather it positions the cursor at the beginning of the next line.

3. CTRL ^ (RS) initiates the recording of the upper half of the screen (8 lines). The 8 lower lines of the screen move into the position of the upper 8 lines, the lower 8 lines will be cleared and the cursor will be at the beginning of line 8.

Before you can do a recording, you must wind the tape cassette manually until the brown oxide coated layer is visible in the center window of the cassette.
(Only with cassettes having a leader and trailer tape).

You then put the cassette into the recorder and press the "start" and the "record" button. The computer will start and stop the recorder after pressing RS.

Playing back your recordings will place them on the screen again starting at the position of the cursor. Do not stop the recorder while the "block" lamp is on.

If the space following the cursor is fewer than 8 lines, the rest of the block is lost.

Most IPS-versions incorporate additional editing keys.

1. **DEL** moves the cursor 1 line up, it thus is the opposite of LINEFEED.

2. **CTRL O (SI)** deletes one character under the cursor, the rest of the line moves one position left and the last character of the line becomes a blank.

3. **CTRL N (SO)** inserts a blank at the cursor. The character under the cursor and the rest of the line moves one position to the right, the last character of the line is lost.

4. **CTRL R (DC2)** (delete line) replaces the rest of the line starting at the cursor by the corresponding part of the next line.
The following lines move one line up, the last line is being filled by blanks.

5. **CTRL Q (DC1)** inserts a line filled with blanks starting at the cursor; the rest of the line following the cursor and all lines below move one line down. The last line is lost.

6. **CTRL X (CAN)** resets the block counter to one. The block counter is incremented with each recorded block and written after RS into the last position of the screen. After count 5 it starts with 1 again. The purpose of this counter is to simplify editing large amounts of text. This concept assumes that your text material will be printed with a hard-copy device onto pages.

If 5 blocks constitute one page (40 lines), block number 5 signals the last block on a page. The block counter thus enables you to keep track of the page format.

If you have completed your text editing, you may leave the edit mode again by pressing CTRL D (EOT). IPS once again expects IPS words starting at line 8.

Instead of manual entries, you may now also play back text recordings. Make sure the cursor is at the beginning of line 8 . Since CR is not recorded, IPS will recognize only complete blocks of 512 bytes.

Entering such a block is equivalent to the CR. IPS will stop the recorder while it is processing the block. After completion it restarts the recorder. If there are further blocks, it will process them one after another.

Because the cassettes contain no inherent file identification, you must write appropriate notes on the cassettes to maintain order in your recordings; there is no automatic file search available due to the limited control the computer has over the recorder.

During program development it is a good idea to not record too many blocks on a cassette, modifications of text are much faster this way. You may overwrite individual blocks on the cassette without problems; you merely have to position the tape to the block gap before the block to be overwritten.

If you listen to the recording, the block gap easily can be identified by a momentary increase in pitch.

You may also record spoken comments on the cassette between blocks; the system will ignore this voice recording.

Later on it will be explained, how you may record data or complete translated programs and how you may enter them again.

3.4 Process control programming

3.4.1 Inputs and outputs

IPS primarily was created to allow for controlling of technical processes. For this purpose there are some fields representing the "outside world". The length of these fields depends on the hardware of your installation, particularly on the number of I/O ports.

The field OUTPUT usually will have four or more bytes. Numbers you store into these field components will appear, bit by bit, on the corresponding output lines of the output-ports. Process control inputs and outputs.

The field INPUT (generally three or more bytes) works in the opposite direction. Signals to the input ports appear in the field INPUT and may be evaluated by reading these bytes. For example:

INPUT 2 + @B fetches the byte at input port 2 and places it on the stack.

```
#55 DUP OUTPUT !B OUTPUT 3 + !B
```

stores #55 (B01010101) into output-ports 0 and 3 and thus forces the output lines to the pattern 01010101 on each port.

The approach of handling I/O ports by fields corresponds to the so-called memory mapped I/O hardware concept. Although all processors may employ this I/O philosophy, processors having special CPU I/O instructions usually lack the supporting hardware of this technique. If your processor has special I/O instructions (COSMAC or

8080), and your hardware does not support mapped I/O, IPS simulates this mode by transporting the contents of the field every 20 ms to the corresponding I/O ports and vice versa. In most applications the resulting delay may be neglected (e.g. a power relay pulls in about 40 ms).

If you cannot tolerate this delay, chapter II will explain how you may design custom I/O without delay.

If your computer contains an A/D converter, there is another field for this input called A/D-CONV . The converter is assumed to convert one analog input voltage (typically 0 - 2 Volts) into a number the computer "understands".

With an 8-bit converter these numbers range from 0 - 255 (one byte, #00 - #FF). The converters used by AMSAT maintain #FF with overranging and #00 with underranging.

Normally there is a commutating switch (multiplexer), which scans a number of analog lines and thus their values are sequentially converted into numbers. These numbers are then placed into the field A/D-CONV .

Each analog input line thus corresponds to a byte in the field and may be inspected by the program.

The IPS operating system controls the multiplexer and puts the conversion results into the field. The converters used by AMSAT convert one level every 20 ms. If there is a large number of analog voltages, say 32, every value in the field is updated every 0.64 seconds. Chapter III contains information for customisation of the A/D converter.

3.4.2 Clock

To simplify programming time-dependent problems, IPS contains a clock supplying the time, day and 4 stopwatches. The field CLOCK supplies time in 6 bytes. (The operating system keeps the time).

CLOCK byte 0 contains 1/100 seconds.
It is updated every 20 ms up to 98, then it is reset and there is a carry into byte 1.

byte 1 contains seconds (0 - 59)
byte 2 contains minutes (0 - 59)
byte 3 contains hours (0 - 23)
byte 4 and
byte 5 contain days, continuous counting without special limit or reset (16-bit number)

You must set the clock after loading IPS, if you wish to use it. For this purpose you may enter the definition:

```
: SET     0 CLOCK ! CLOCK 2 + DUP 2 +
             EACH I !B
             NOW       0 CLOCK 5 + !B ;
```

You then may enter **15 12 24 SET** and the clock is set to date 15, time 12:24 hours.

If your program needs to check the time for a limit, it is recommended to proceed backwards from date to hours and so on. You thus avoid problems if the time changes during your checking. The following definition checks if a certain time has been reached or exceeded. If not, it will deliver 0, else 1 onto the stack.

```
: YET SWAP CLOCK 4 + @B >     ( HOURS NOT YET REACHED )
       YES? DEL 0
       NO:  CLOCK 3 + @B > ( MINUTES NOT YET REACHED )
              YES? 0
              NO:   1
              THEN
       THEN ;
```

e.g. 12 30 YET delivers 0 before 12:30 and and 1 afterwards.

A more simple test is possible, if your time limit is available in a field. In this case you may use F-COMP to compare the clock against your limit and immediately get the result of the comparison. (During F-COMP and >>> the clock does not change).

3.4.3 Stopwatches

In addition to the clock there are four stopwatches. They are fields, 4 bytes each, and are named SW0 , SW1 , SW2 and SW3 .

SW0, 1, 2, 3	byte 0	contains 1/100 seconds	(98 - 0)
	byte 1	contains seconds	(59 - 0)
	byte 2	and	
	byte 3	contain minutes	(16 bits, 32767 max)

Byte 0 is reduced by 2 every 20 ms by the operating system. If byte 0 contains an odd number, the stopwatch is stopped. You may thus stop it momentarily by incrementing byte 0. You may later restart the watch by reducing byte 0 again by 1.

If the time of the stopwatch has expired, all bytes are 0 except byte 0; it is 1. To test whether the stopwatch has expired, you only need to test byte 0. While the watch is running, this byte is even and thus will fail the YES? test.

e.g.:

```
SW1 @B YES?      ...      ( WATCH EXPIRED )
       NO:       ...      ( WATCH STILL RUNNING )
       THEN
```

In setting the stopwatches to a particular time, byte 0 should be the last byte set; the watch will start only then. (Or you may use >>> to set all four bytes at a time).

e.g.: (SW2 is to run for five minutes)

```
5 SW2 2 + ! 0 SW2 !
```
(byte 0 and 1 are set together, which is OK)

If you are interested whether a stopwatch is below a certain value, you again may use F-COMP . The stopwatch is not updated while F-COMP is being executed.

3.5 Repeated execution and multiprogramming

3.5.1 The chain

So far we learned only how to execute a definition once by typing its name.

Many technical problems require a continuous repetition of a program. A routine checking a machine for example, is to perform this check periodically; in case of problems it then is to perform the appropriate actions.

In the general situation of the computer controlling outside activities, there are usually a few more or less independent processes the computer is to look after.

This multiprogramming and the repeated execution is solved with IPS by the concept of a chain. This is a field containing not numbers, but programs that are to be performed in sequence. IPS executes these programs one after the other, and at the end of the chain starts again at the beginning.

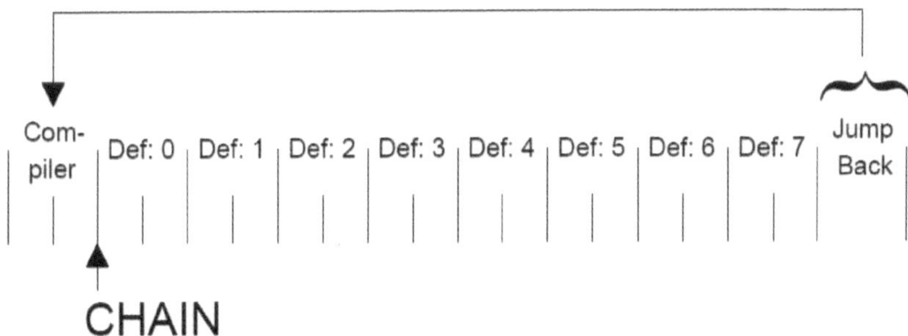

The chain has eight positions in which to put programs. A ninth position in front of the chain positions contains the COMPILER, the program processing keyboard and tape inputs.
Position 0 contains the program displaying the stack contents; all other positions are preset with no-ops, words resulting in no action of the computer. You may add further programs to the chain which then are also periodically executed. Experience shows

that even with large programs it seldom takes longer than 100 or 200 ms to run once through the complete chain.

For many technical applications this is fast enough and eliminates the need of interrupts and their associated problems typical with real-time programming.

To put or remove programs from the chain is very easy; it may take place by keyboard or by definitions. In particular, a definition may remove itself from the chain.

It later may be put into the chain again by another definition. With these operations the rule always holds that a definition in the chain is always executed to the end. Its removal from the chain means only that during the next pass through the chain the definition no longer exists in the chain.

Before you put definitions into the chain, you have to make sure that they leave the stack as they find it upon entry. A definition leaving a number on the stack will leave this number upon each pass of the chain; thus the stack will overflow or underflow resulting in a program crash.
(While executing definitions in the chain, the stack is not checked.)

To put a program into the chain, you first specify the chain position (0 - 7), followed by the word ICHN and the name of the program to be put into the chain.

e.g.:

4 ICHN TEST	puts the definition TEST into position 4 for periodic execution.
4 DCHN	removes the definition at position 4 and replaces it by a no-op.

In particular, 0 DCHN removes the stack display from the chain. This may be useful if you have other ideas on the use of the display. To switch it on, you may type 0 ICHN STACKDISPLAY .

Of course you might have put the stack display into position 5 as well, the net result would have been the same.

If you design programs in- and de-chaining each other make sure that there are no states from which there does not exist an exit. It is a good idea to prepare a list of the use of the chain positions and the conditions under which definitions are put there and removed.

Although the chain is a simple tool to allow multiprogramming, there may exist situations where this concept is not fast enough. In chapter III another concept is presented allowing high-level interrupts (so-called pseudo-interrupts) to exist in parallel to the chain.

Where even this is not fast enough, the assembler of chapter II will enable you to use machine interrupts. The two later techniques are slightly more complex to use, thus the chain is the preferred mode if the timing is not critical.

The following table relates the applicability and response time requirements of the three techniques.

a. Chain:	Up to a few 100 milliseconds until service depending on rest of chain.
b. Pseudo-interrupt:	Up to 18 ms until service with single user-pseudo-interrupt. With more, add duration of higher priority routines.
c. Machine interrupt:	50 - 250 µs until service depending on processor and hardware.

3.5.2 Waiting for events in the chain

A common situation in a program is: can you proceed with execution only if certain internal or external events have occurred? In principle you could program such parts using the LBEGIN ... LEND? construct for waiting. But the computer then "hangs" at this point and cannot execute other programs in the chain in between.

There are two words to solve this problem. The idea behind them is to remove from the chain the program which cannot be continued at the present time and to replace it by a program which checks on each pass whether or not the required event has occurred. If so, the original program is again put into the chain and resumed at the position after the suspension. To make this possible, IPS stores at such a wait call the original chain content and the position of where to resume in it later.

The word WAITDEF is used to effect the suspension; it expects on the stack the address of the definition to be used to do the wait and checking. The word 'followed by the name delivers the address put by WAITDEF into the chain. The action is reversed by the word YES:CONT . This word expects a 1 or 0 on the stack. With a 0 no action results, with a 1 the original program is again put into the chain and continued after the WAITDEF .

The following example assumes that stopwatch SW3 is available for waiting.

The program is to continue only after it has waited for a specific time. The actual waiting is done by the definition TIMEOUT .

```
: TIMEOUT SW3 @ YES:CONT ;
```

If the stopwatch has run out, the least-significant byte becomes odd activating the YES:CONT and thus resuming the program. We still need the definition which initiates the waiting.

```
: M/WAIT SW3 2 + ! 0 SW3 ! ' TIMEOUT WAITDEF ;
```

This definition expects a number on the stack; the number of minutes to wait.

By calling it, the program will continue after this number of minutes. For example, the MAINPROG in the chain is to wait for 20 minutes before continuing:

```
: MAINPROG ... 20 M/WAIT ... ;
```

While MAINPROG waits for the 20 minutes to expire, the rest of the programs in the chain are periodically executed.

IPS has to store some addresses to enable the WAITDEF and YES:CONT.

This results in the restriction that YES:CONT may only occur in the definition that is directly put into the chain. (Since only one return address is stored the YES:CONT should not be invoked by further nested definitions).

The WAITDEF has a similar restriction, only it is to occur in a definition which is called by the definition in the chain.

Furthermore, while invoking such a wait action, the stacks should be empty (do not use it inside a EACH .. NOW). Routines containing wait actions can only be tested by putting them into the chain. Do not call them by the keyboard.

3.5.3 Limited input mode

With IPS the keyboard and the display may be used for two different purposes. Initially they serve to help the program development and debugging, later they may be used to operate a machine or other device. In this case the program must be protected from inadvertent abuse. This can be achieved by entering the "limited input mode".

The effect of this mode is to allow access to only certain definitions – those which are required to control the application. All other words are marked by ? and not executed.

Furthermore, numbers are not recognized and are not put on the stack. Rather they will be accepted only if they were requested by the program and then are put into the variable N-INPUT . While the program is waiting for a number input, no other words are recognized.

The words to be recognized must be the last words defined into the dictionary. Entering the limited input mode involves two steps. The typecode address (' name) of the earliest word in the dictionary to be recognized must be stored in the variable LIS . For instance, if OFF is the first word to be recognized, type:

```
' OFF LIS !
```

After this store a 1 into the variable LIM (limited input mode) .

```
1 LIM !
```

puts IPS into the limited input mode and it will then only recognize OFF and entries defined thereafter.

If you desire to request a number from the computer operator, set the variable N-READ to 1 . Your program then needs to check if N-READ again has become 0; in this case the requested number is available in the variable N-INPUT . The actual input resets N-READ . Until a valid number has been entered, no other words are recognized.

You may leave the limited input mode by typing RUMPELSTILZCHEN .
This resets LIM and all words are again available.

4 IPS - Operation and memory organization

By now you have learned all important properties of IPS except how to record and re-enter data and translated programs by tape. In principle you could
enter all programs always as source, but this requires more time and tape than entering objects in their translated form.

To understand the necessary actions, let us have a look at the IPS-memory organization first. This is only a cursory introduction, the details are presented in chapter III.

4.1 The address space

A typical computer on which IPS is running is assumed to have 16 Kbytes RAM. These generally would be located starting at address #0000 and extending up to #3FFF (0 - 16383).

In addition 1 Kbyte of memory is represented by the TV-screen. This segment may start in address space either at #8C00 or at #CC00.

The display electronics treat the first seven bits of each byte as an ASCII-character and displays it on the screen. The most significant bit of each byte is used to invert character and background brightness - the cursor is generated this way.

The video electronics continuously scan the 1 Kbyte memory to produce the displayed picture. If the computer needs access to this storage though, it has priority. The display is blanked for this duration. Because these accesses are very fast, this is hardly noticeable to the observer.

With frequent accesses you may notice some short blanked lines running over the screen.

Earlier you learned that IPS uses a dictionary, a nametable and two stacks.

The dictionary grows linearly in memory starting at lower addresses and extends to about #1400 depending on the particular version. HERE delivers the first unused position of the memory.

The first pages (1 page = 256 bytes) contain machine code to connect IPS to your processor, the rest is IPS code constituting the system and a buffer for tape output (512 bytes).

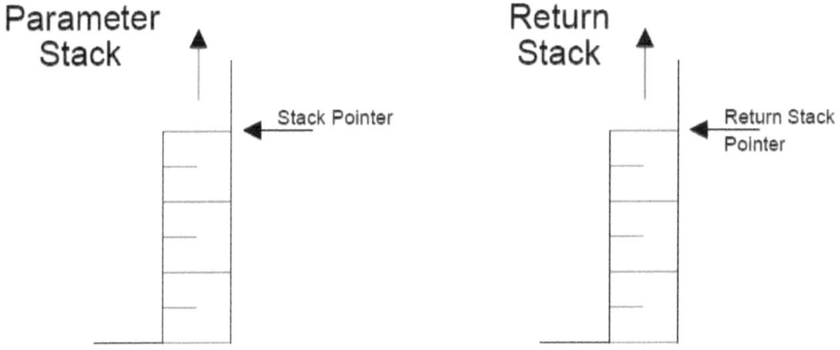

The nametable and the stacks fill the memory from the other end downwards.

The nametable is located between #3400 and #3FFF. except for the 6502, the return stack extends from #3400 downwards and the parameter stack starts #3300 downwards. (6502: return stack #0200, parameter stack #3400) IPS has two variables containing pointers to the last position in use by the stack. Thus the system "knows" how many numbers it has on the stack; it keeps track of the entries by decrementing or incrementing these pointers. Only 16-bit numbers are held on the stacks. Thus the stack pointers always change in twos.

4.2 The dictionary

Each dictionary entry consists of an entry in the nametable having six bytes. The first four contain a code for the name (length of the word and a 3-byte code) and a pointer to where the entry proper starts in memory. The entry itself always starts with a typecode identifier (a pointer to executable code) and additional parameters.

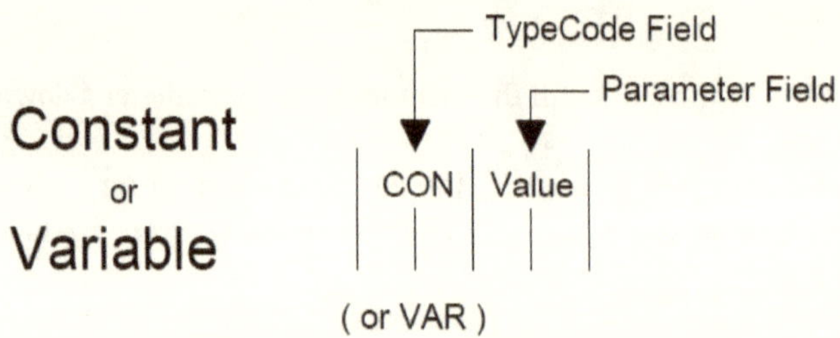

With variables and constants there is a two byte parameter field containing the actual value.

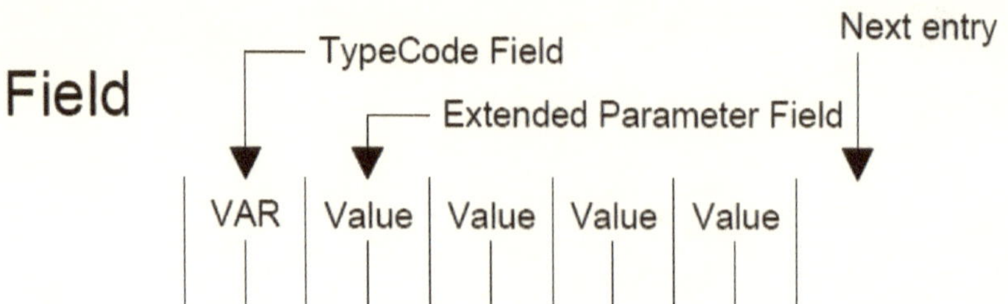

Field entries have an extended parameter field to maintain additional numbers, but otherwise are identical to variable entries.

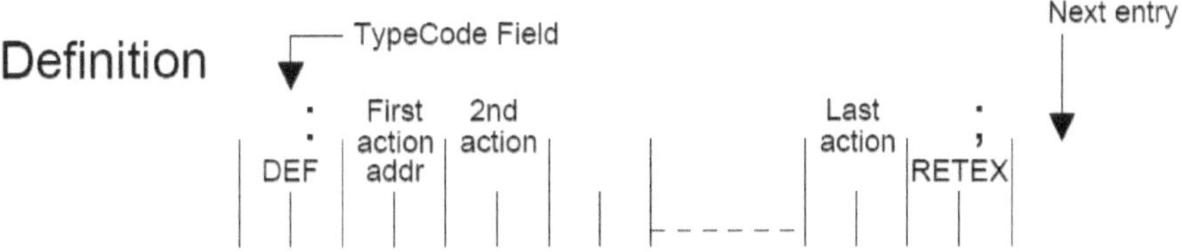

Definitions contain, in the parameter field, the addresses pointing to the typecode of the particular entries constituting the definition. Because these addresses are compiled into the definitions rather than the names, IPS executes extremely fast. There is no search action necessary, after fetching two pointers the position of executable code is available.

Most entries into a definition result in two bytes of code. But there are some exceptions. Numbers require 3 or 4 bytes depending on whether they are between 0 and 255 or not.

If certain numbers are required often in a program, it thus may be of advantage to create a constant entry. Large numbers used more than 5 times or small numbers used more than 10 times justify a constant entry. Referencing a constant in a definition requires only two bytes.

Entries connected with branches do not result in a straightforward 2-byte address compilation. These words are really executed during the compilation process to handle jump addresses. These are maintained on the stack during compilation and explain the numbers seen on the stack during a definition compilation. The list shows the number of bytes compiled with the jump words.

```
ICHN          6 bytes      NO:            4 bytes
DCHN          6 bytes      THEN           0 bytes
EACH          4 bytes      LBEGIN         0 bytes
NOW, +NOW     4 bytes      LEND?          4 bytes
YES?          4 bytes      THEN/REPEAT    4 bytes
```

All entries eventually must lead to executable code. These entries are called RCODE or CODE. The entry simply consists of a pointer pointing to actual machine code which the processor can execute. The IPS code routines are collected on the lower pages of memory to simplify transcription of the system to new processors.

Thus the RCODE entries point to this code. CODE entries produced by the assembler have the executable code following the entry.

4.3 The emulator (inner interpreter)

IPS words are executed in three stages. There is a Pseudo-Program Counter (PPC) pointing to the next word to be emulated. The content of this location is read and is called the Header Pointer (HP). The PPC is incremented by two to point to the next instruction in line. HP points to the typecode field of the entry to be executed.

The content of this position is read; it points to executable code. The emulator jumps to this position to execute the code there. After this execution the code contains a jump back to the emulator and the emulation cycle starts all over again.

If the emulator executes a definition, the code (called DEFEX) executed at the beginning of the definition places the PPC on the return stack. HP+2 becomes the new PPC.

This sub-routine calling mechanism does not require an explicit jump address; it is implicitly available by the indirect emulation.

The emulator requires only a few instructions and a few bytes of code; with some processors it is faster than a standard sub-routine call and thus responsible for the fast execution of IPS.

4.4 The 20 ms pseudo-interrupt

The emulator is designed in such a way as to accept interruptions between the execution of code routines. This is similar to conventional interrupts, only the high-level stack-computer is interrupted resulting in no "saving overhead".

Every 20 ms such a pseudo-interrupt is forced to do various tasks that have to be executed periodically. Most important, the clocks have to be updated and with some processors I/O actions are performed.

The pseudo-interrupt principle has two important consequences.

First, the latency of the interrupt is governed by the duration of the longest code routine; under no circumstances should code routines take longer than 15 ms to execute.
(This is easy to ensure with modern processors)

Second, since code routines are not interrupted, it is assured that during the execution of the code the interrupt related activities do not take place. Thus, the time of the clocks does not change, e.g. while the code >>> is being executed.

The same applies with I/O;
if >>> is used, all outputs will occur within microseconds at the same time. You need not fear that part of your output will show up 20 ms later.

4.5 The compiler

The compiler is the program processing keyboard and tape inputs. It is entirely written in IPS and thus not directly related to the low-level activities discussed so far. If there is an input word, the following actions are taken:

1. The compiler checks whether the word is available in the nametable. Depending on the present mode, the word is either executed or compiled.

2. If the word is not in the nametable, the compiler checks whether it is a number. In this case it is either put on the stack or compiled into the definition.

3. If the word is neither a known name or valid number, it is marked by an error ? and further input processing is halted.

5 Cassette tape operations on binary data

5.1 Data storage

To facilitate recording data on cassettes IPS maintains a field named $BU having a capacity of 512 bytes.

You may write any data into this field and then record them by typing RECORD . For the duration of the recording you should not write into the buffer, of course.

You may check by program if the recording is still taking place by calling the variable C/Z (cassette counter). This variable is only one byte wide and thus has to be fetched by C/Z @B .
During the recording action it is greater than zero.

5.2 Reading data cassettes

You may read data into the buffer $BU again by typing READ . While the READ takes place, the variable LOADFLAG (1 byte) is 1. After the completion of the reading (512 bytes being put into the buffer) the system resets LOADFLAG to 0 again. During the READ the keyboard may only be used to write to the TV.

The motor of the cassette recorder may be controlled automatically. It is switched on if the variable LOADFLAG is 1 or if the variable TEXTREAD is 1.

Because TEXTREAD is initialized by IPS to 1, the motor is always switched on except for the time IPS is processing text input. An input error resets TEXTREAD thus inhibiting further reading. The word OK sets it again.

If you want to enter data only, you must set TEXTREAD to 0 to prevent data from being treated as text input. Then the recorder is started only if a READ has requested data. The variables LOADFLAG and TEXTREAD, like C/Z , are 1-byte variables. Use @B and !B only.

If you design your own buffers (e.g. a cyclic buffer), you may read data into these buffers directly by using the word $LOAD . $LOAD expects the address of where the data is to be loaded and on top of it the address of the last byte to be loaded plus one.

Thus the definition READ is defined as

```
: READ $BU $BU 512 + $LOAD ;
```

Note that this way you may enter multiple blocks, but you cannot enter fractional blocks.

5.3 Program storage

Programs are stored by using the program DUMPER (appendix A.10). You first enter this program (3 blocks) like any other IPS program. It requires 74 bytes of storage still left above $H and the translation process will leave one number on the stack. It does not add any names to the nametable.

Now type $TUE and your program including IPS is recorded as a whole on the cassette.

You re-enter this program into the computer simply by loading the tape just like the IPS tape, only your load now contains IPS plus your program.

By this means you also may duplicate IPS. Note that with the program DUMPER you can produce only one copy; by doing the copy, DUMPER is destroyed. You thus have to re-enter DUMPER for each copy.

6 IPS programming hints and exercises

6.1 Programming method

To write IPS programs proceed in the following order:

1. Define the problem by a description or a flowchart.
   Also define the data structures and how you want to organize them.

2. Now write down definitions describing the problem
   by words having specific intended functions.

3. These specific function words are defined in terms of new words
   describing in more detail the actions of the words.

4. Continuing this process you work downwards
   until all details are described using only IPS words.

5. Now the definitions are entered, the lowest ones first,
   and tested by typing their name.
   If they work properly, you enter the next higher level and test these definitions.
   Eventually all definitions are entered and tested, the program is ready.

Note that you start writing a program by always defining the more general items first (top-down design), but you must enter it in reverse since you can reference only those items already defined.

It is not a good idea to start writing a program by defining details first which you feel might be useful later. You lose only time this way because these details never turn out the way you need them later.

If you have already programmed in FORTRAN or BASIC, try to resist the temptation to name too many variables. As a guideline, use the rule that variables should be named only if you must access them from more than one definition (global variables).

If you need a variable only for in-between storage within a definition, use the stack or the return stack.

You may remove items from the dictionary again and thus reclaim the memory space by typing DEL/FROM name . For instance typing DEL/FROM TEST removes the word TEST and all words defined after TEST from the dictionary and the nametable.

When you create definitions, it is a good idea to adopt the same philosophy with regard to the stack as IPS uses itself.

This means that definitions remove all parameters intended for it and leave only explicit results. This helps you to avoid hard to locate stack errors (unaccounted items on the stack).

Also make sure that both ways through a YES? ... THEN leave the stack with an identical number of parameters on it.

If you have problems with a definition, simply test it by keying the words constituting it individually and observe the stack. Most errors are readily located this way. Of course, you must manually simulate the effect of branches and loops.

Do not test too many untried definitions at once. Debugging may get too involved this way. IPS offers you the possibility to start with the low-level items first. You thus are able to create a base of proven definitions on which you can rely at the next higher level.

Before you put definitions into the chain, make sure beyond any doubt that they leave the stack as they found it. If you have definitions in the chain, first dechain them prior to eliminating them by DEL/FROM .

6.2 Exercises

The following exercises were designed to acquaint you with the use of the more common IPS words. The solutions given are by no means the only possible solutions and possibly not the best. Type in your solutions and give them a test.
You will learn more this way than by hours of theoretical deliberations.

a. Write a definition adding all numbers from 1 to N, N being supplied on the stack. Leave the result on the stack.

b. Modify a. in such a way that it sums M to N, both on the stack.

c. Modify b. in such a way that it will accept M and N also, if their order on the stack is reversed.

d. Modify b. in such a way that it will add only the even numbers of the sequence.

e. Modify b. such that it will add only the numbers divisible by seven.

f. Write b. in such a way as to use the other two repetition constructs.

g. Write a definition filling 4 TV-lines starting at TV4 with all the characters your display is capable of (one byte may contain 256 different numbers and thus characters.)

h. Write a clock displaying the time HH:MM:SS at TV4. Update it once per second. (This exercise is more difficult; you may need a few test iterations until it works properly)

Note, that using CONV the result may have only one digit.

Make sure that this does not spoil the display.

i. Write a hex dump memory display program.

6.3 Solutions to the exercises

```
a. : 1SUM 0 1 RTU RTU EACH I + NOW ;
b. : SUM 0 RTU                     EACH I + NOW ;
c. : SSUM SOT SOT > YES? SWAP THEN SUM ;
d. : ESUM SOT YES? ( ODD ) SWAP 1 + SWAP
                  THEN 0 RTU EACH I + 2 +NOW ;
```

(This solution is preferable to an even/odd test within the loop. Why?)

```
e. : 7SUM 0 RTU EACH I 7 MOD =0 YES? I +
                                    THEN
                        NOW ;
```

(Also try to write it using the technique of d.)

```
f. : SUM 0 S>R SWAP SOT SOT >=
           YES? LBEGIN DUP R>S + S>R 1 +
                  SOT SOT <
                        LEND?
                             THEN DEL DEL R>S ;

     : SUM 0 S>R SWAP LBEGIN SOT SOT >=
                      YES? DUP R>S + S>R 1 +
                      THEN/AGAIN DEL DEL R>S ;

g. : DISPLAY TV4 TV4 255 + EACH I I !B
           ( !B STORES ONLY LOWER BYTE OF I )
                          NOW ;

h.
: TIMEWRITE SP @ " 00:00:00 " SP ! CLOCK 3 +
    0 2 EACH DUP @B DUP 10 < YES? SP INCR
                                 THEN
            CONV SP INCR 1 -
       NOW DEL ;
```

```
0 VAR AUX
: TDISPLAY CLOCK 1 + AUX 1 F-COMP INV
    YES? TV4 SP ! 10 BASE ! CLOCK 1 + AUX 1 >>>
           TIMEWRITE
       THEN ;
2 ICHN TDISPLAY
```

i. A program displaying a page (256 bytes) of memory as a list of hexadecimal numbers.

```
( HEX-DISPLAY FOR IPS-N )
: NUMB SWAP #30 + DUP #39 >         ( WRITE No. in HEX )
                 YES? 7 +
                 THEN SOT !B 1 + ( POS ) ;

: NUMBI NUMB 1 - DUP @B #80 OR SWAP !B ;
                               ( WRITE INVERTED No.)

: POX
 R>S I #40 * I #F0 AND 4 / + #3FF AND TV0 + SWAP S>R ;

: HEXDISPLAY
   0 255 EACH DUP I + @B DUP #F AND SWAP 16 / POX
             NUMB NUMB #2020 SWAP !
         NOW WEG
   0 15 EACH I I 4 * 2 + TV0 + NUMBI       ( ANNOTATE )
             I I 64 * 63 + TV0 + NUMBI     ( DISPLAY )
         NOW
   TV8 2 + $CEN 0 $P2 ! ;

: 0H    0 HEXDISPLAY ;

: 1H 256 HEXDISPLAY ;
```

After translation the program is used by typing 0H or 1H . It then displays page 0 or page 1 of the tape buffer using the full TV-screen in a column format.

An arbitrary section of memory may be displayed by typing the starting address followed by HEXDISPLAY .

After 0H , 1H or HEXDISPLAY the stack display is momentarily disabled to prevent clobbering the display.

The display is annotated with horizontal and vertical 0 - F in inverse video characters.

7 A larger sample program (StarTrek)

This sample program was modelled after the StarTrek described by David Price (Byte Vol.2, No.3 (March 1977), pp 106). The main differences are in the command structure and in the fact, that the Klingons fight back. The original version was written in BASIC requiring 22 Kbytes of memory. The IPS version uses about 11 Kbytes including IPS.

[At this point the example program should follow but, tantalisingly, has not been found either in paper or electronic form. - Ed.]

Where is the machine readable source of the infamous StarTrek program?

Kirk: Has Meinzer lost it?

Spock: Logical Captain; illogical Meinzer! [Gape,1983]

NOTE Juergen Pintaske:

If anybody can supply this missing part, I will try to include it here
- or at least the link to where this part can be found for download.

Please send to epldfpga@aol.com

42

CHAPTER II

The Assemblers

1 General

IPS is sufficient to program almost all problems without having to know the structure or machine language of your processor.

But there are three exceptions:

a. You want to interface to some special hardware. The IPS I/O structure may not be sufficient for this purpose, particularly if interrupts are involved.

b. You want to program an extremely time critical problem, that cannot tolerate any overhead.
The IPS emulator has an overhead of about 40 – 50 µs per executed word.
Further, IPS treats all numbers as 16-bit quantities.
This results in additional overhead if you are mostly concerned with 8-bit quantities. The combined effect results in IPS-programs running about 2 - 3 times slower than optimum machine code.

c. The mathematical precision available from 16-bit integers is not sufficient for your problem and you want to add multiple precision or special arithmetic operators.

All these problems are solved elegantly by being able to define words, not in terms of other IPS words, but by machine instructions.

These words then may be used like all other IPS words, but by being defined at the machine level they give you the facilities mentioned above. The program allowing you to add the machine instruction definitions is called assembler.

To use it you must of course be familiar with the instruction set of your processor.

The IPS assemblers are intended mainly to allow you to build short interface routines between your processor, your hardware and high-level IPS. They are not intended for extended programs.
It is not reasonable to do any extended assembly programming with IPS because assembly programs are harder to debug, usually require more memory and most important, prevent your programs from being run on other computers using a different processor.

It is thus a good idea to keep the machine routines as short and simple as possible.

2 Basic philosophy of the assemblers

The IPS assemblers create assembly language definitions by using the word CODE followed by a name instead of using the colon. The last word is NEXT instead of the semicolon.

In contrast to the creation of high-level IPS definitions, the assembly takes place in the keyboard mode; definition mode is not entered.

Typing a mnemonic, e.g. LDA , results in the appropriate code being deposited at HERE and then HERE being incremented to the next free position. This has important ramifications as we shall see later.

There are two major departures from conventional assemblers.

First, the sequence of specifying operations follows the IPS convention, i.e. first source, then destination and then the operation. Most conventional assemblers use the reverse order.

Second, the usual branch mnemonics are not used, rather the words Y? N: TH and BEGIN END are used similarly as in IPS. The Y? and END are preceded by a condition (usually one of the condition codes of the processor status register). Thus, there are no labels or GOTOs used with the assemblers.

Just as with IPS, the stack is used to handle the jump addresses. Let us have a closer look at this mechanism. If the word Y? is encountered, the appropriate jump code is deposited at HERE and the address of the following position is put on the stack; HERE is then incremented further to where the next instruction is to be placed. Thus there is on the stack the address where later the jump destination of the Y? is to be inserted.

When the TH is encountered later, the Y? jump may be completed; HERE is the address where the jump is to go. Thus, TH takes the address off the stack left by the Y? and inserts there the jump destination.

The N: similarly assembles an unconditional branch leaving on the stack the position where the jump address is to be inserted. The address left by the Y? is removed and used to insert the position following the N: jump into the Y? .

The BEGIN is equivalent to HERE ; it simply leaves the address to which the END must assemble a conditional branch. Because the addresses are on the stack, you may manipulate them, e.g. by SWAP, to be able to depart from the strict IPS nesting rules or DUP them to return from multiple points to a loop begin.

The assembler works in the interpretive mode. Thus you may use any regular IPS words in between assembly mnemonics. (e.g. to compute numbers used in immediate instructions) You may also use the word NEXT as often as you wish to create multiple exit points from a code definition.

You may collect a recurring sequence of mnemonics into a regular definition; calling it each time deposits the whole sequence of assembly instructions.

Such definitions are called macros. Macros of course may contain the structuring words (e.g. Y? or END) as well.

Because IPS uses different structuring words, you might use these as well to create different assemblies depending on parameters received by the macro definition. This facility is called conditional macro assembly.

Although IPS offers these advanced capabilities, do not get carried away by them. Always remember that assembly routines serve a sort of crutch function and should not be allowed to mushroom. Even in extremely time critical situations it usually pays only to code the innermost loop in assembly language.

The effect of the high-level overhead on total execution time is negligible if outer loops are coded in IPS.

The main function of machine code routines is to create an interface to IPS via the stack. At the machine level the stack operations are not automatic.

Depending on the processor, the stack is maintained explicitly by a pointer and appropriate machine code instructions. If the processor already has instructions for stack operations, this stack is usually the parameter stack.

The return stack is simulated by another pointer. (exception: 6502)

The assemblers by and large use the original mnemonics of the manufacturer.
Unfortunately, this approach precludes much commonality between the several processor's assemblers. But most likely you will work only with one processor, anyway. Thus, the differences will be of little consequence.

IPS occupies certain machine resources. Make sure that you are familiar with the registers or memory sections used by IPS, which are not to be disturbed.

Handling interrupts is a bit more involved and thus is postponed until chapter III.

The assemblers are written in high-level IPS. Thus, they may also be used to produce code for different processors than the one on which they are running.

This cross-assembly is particular useful for transplanting IPS onto new processors.

In addition to the mnemonics for the individual processors all assemblers allow to deposit in-line numbers by using the word , . One byte is deposited at a time.

All assemblers may create certain error messages. These differ from processor to processor but are self explanatory. If you encounter such a message, generally it is not recommended to manually correct the code. Rather the entire code word should be re-assembled because the stack may have been modified unpredictably.

If your assembly leaves numbers on the stack or results in "stack empty", you may take this as an indication of an error in your code. Do not execute this code, rather re-assemble it word by word to identify the problem.

The 8080 assembler is coded using English IPS words; the others use the German set.

3 The CDP 1801 assembler

AMSAT makes use of numerous CDP 1801s. This assembler therefore recognizes only the instruction subset of this processor. If you wish to make use of the extended instruction set of the CDP 1802, it is a relatively trivial exercise to add these to the assembler. (See section 7).

3.1 Register allocations

R0 is the DMA pointer used for recording of cassettes
R1 contains the entry address of interrupt program handling cassette inputs
R2 also named RS is the return stack pointer pointing to the lowest address n use. Decrement before storing into it !
R3 is the program counter during the execution of all code routines
R4 used for temporary storage like RD, RE and RF
R5 unused by IPS
R6 is the program counter during the low-level interpretation of IPS words. (Emulator pc)
R7 is the pseudo-program counter (PPC) of IPS
R8 input pointer for tape inputs while LOADFLAG is set to 1
R9 input pointer for keyboard and tape input with LOADFLAG =0
RA system page (#04) pointer. RA.1 should always contain #04
RB also named PS is the parameter stack pointer. It points to the lowest address in use. The parameter stack always handles two-byte numbers, the lower address containing the more significant byte. Do not use RB from interrupt programs because code routines may temporarily store relevant data up to 8 bytes below pointer position.
RC is used as the emulator header pointer + 2 (HP+2). If you do not need this information in your code, you may reuse it and destroy its contents.
RD used for temporary storage within code routines
RE you may use them without restriction. But they cannot (RF) be used for storage beyond your code routine.

The X register is not being kept pointing to any particular register. You must set it in each code routine prior to using the first X referencing instruction.

Between code routines X may get changed by the operating system, so never forget to set X ! .

EF 1 is connected to the 20 ms flip-flop, it is reset by output 5.

EF 4 is the keyboard input, it is scanned every 20 ms.

If EF 4 is active, output 6 instructions are used to strobe in 7 bits of the character sequentially via EF 4, LSB first.

Outputs 5 and 7 are reserved for special functions; see chap. I, sect 3.4

3.2 The instruction mnemonics

The following mnemonics are identical to the RCA mnemonics; they expect a number (0 - 15) on the stack.

You may also use the constants RS (2) or PS (#B) instead:

```
INC  DEC  GLO  GHI  PLO  PHI  LDA  STR
SEX   and SEP are replaced by ->X ->P
INP and   OUT are replaced by I/O
     ( 0-7 output, 8-#F input )
```

The instructions IDL , RET , DIS , SAV and SHR are identical to RCA usage and do not expect a parameter.

The ALU instructions are changed, the accumulator is renamed A.

LD , ORA , AND , XRA , ADD, -A , A-

are followed either by MX or IM

(separate word). If it is IM, a number is expected on the stack of which the lower byte is used in the immediate instruction.

(e.g. #20 LD IM)

Control structures are created by: Y? N: TH and BEGIN END .

The words Y? and END must be preceded by the conditionals:

```
D=0 , DF , n ( 1-4 ) EF or 0
```

(the 0 has the meaning of never).

The conditionals may be followed by NOT to express the opposite.

(e.g. 2 EF NOT Y?)

Also the word SKP is available.

3.3 An 1801 assembler example

This code routine checks EF 3 and returns a 1 on the stack, if EF 3 is set, otherwise a 0.

```
CODE EF3       1 LD IM  3 EF NOT Y? SHR
                                 TH
               PS DEC PS STR SHR PS DEC PS STR NEXT
```

3.4 Listing of the CDP 1801 Assembler

```
:INT CODE ENTRYSETUP JA? HIER VERT !
                    DANN ;
: ,       HIER !B $H INCR ;
16 FELD $JERROR " DESTIN. OFF PAGE " $JERROR !T
16 FELD $CERROR " PREINSTR. ERROR! " $CERROR !T
: $INADR ZWO ZWO EXO #FF00 UND
      =0 JA?    VERT !B
         NEIN: WEG $JERROR SYSWRITE
         DANN ;
: $TERM WEG $CERROR SYSWRITE ;
: $AD VERT DUP 15 > ZWO <0 ODER JA?    $TERM
                              NEIN: + ,
                              DANN ;
: $ALUACT ZWO DUP 7 > VERT 6 = ODER
         JA?    $TERM
         NEIN: $AD
         DANN ;
:  INC #10 $AD ; :  DEC #20 $AD ; :  RET   #70    ,   ;
:  GLO #80 $AD ; :  GHI #90 $AD ; :  SAV   #78    ,   ;
:  PLO #A0 $AD ; :  PHI #B0 $AD ; :  DIS   #71    ,   ;
:  LDA #40 $AD ; :  STR #50 $AD ; :  SHR   #F6    ,   ;
:  ->X #E0 $AD ; :  ->P #D0 $AD ; :  IDL   #00    ,   ;
:  I/O #60 $AD ; :   NEXT #D6 , ; :  SKP   #38    ,   ;
:  MX #F0 $ALUACT ; : IM #F8 $ALUACT ,   ;

0 KON LD            1   KON OR    2 KON AND     3   KON XOR
4 KON ADD           5   KON -D    7 KON D-     10   KON D=0
11 KON DF           2   KON RS    11 KON PS
:    NOT      8 - ;
:    EF      11 + ;
:    BEGIN       HIER ;
:    Y?      #30 $AD HIER $H INCR ;
:    N:      #30 , HIER VERT $H INCR HIER $INADR ;
:    TH      HIER $INADR ;
:    END      #30 $AD HIER VERT $INADR $H INCR ;
                ( END CDP 1801 ASSEMBLER )
```

4 The 8080 assembler

4.1 Resource allocations

Since the 8080 does not contain enough registers on the chip to maintain all pointers required by IPS, most of them are located on the system page (#300).

Only the pseudo-program counter is maintained in the register pair B,C (also named PC).
The stack pointer on the chip is used for the parameter stack; the return stack is simulated with a pointer stored on the system page.

4.2 The instruction mnemonics

The following mnemonics are identical to the original 8080 mnemonics with the exception that the additional parameters precede the mnemonics as separate words:

```
MOV , ADD , ADC  , SUB  , SBB  , ANA  , XRA  , ORA  ,
CMP , RST , POP  , OUT  , IN   , XTHL , XCHG , EI   ,
DI  , HLT , INR  , DCR  , LDA  , MVI  , LXI  , STAX ,
INX , DAD , LDAX , DCX  , PUSH , NOP  , RLC  , RRC  ,
RAL , RAR , DAA  , CMA  , STC  , CMC  , SHLD , LHLD ,
STA , LDA , SPHL , PCHL
```

Except for MVI and LXI the immediate instructions do not use separate mnemonics, but use the general mnemonics.
These are preceded by the number followed by the word IM . (e.g. **#41 IM XRA , but #41 E MVI**)

The 8080 incorporates unconditional and conditional call and return instructions.
The unconditional cases use the words SCALL (preceded by the address) and SRET .
The conditional cases use the words CALL and RET . CALL needs an address followed by a condition code (see further down); RET is only preceded by a condition code.

With registers a clear distinction is made between 8-bit registers and register pairs. The 8-bit registers are identified by the letters B C D E H L and A , M being used for memory references using H,L as pointer.

Register pairs are identified by the words B,C D,E H,L and PSW . The stack pointer is referenced as $SP to avoid confusion with the IPS screen pointer SP. B,C
may also be referenced as PC ; remember to save these registers (e.g. push them on the stack) if you wish to use them.

Example: **`#400 D,E LXI or H,L INX , but L INR`**

Control structures are created by words Y? .. N: .. TH and BEGIN .. END .

Y? and END must be preceded by one of the following condition codes:

`NZ    Z    NCY    CY    PO    PE    POS    NEG    or    NEVER`

The same codes are used by the CALL and RET instructions.

4.3 An 8080 assembler example

This code routine transfers a byte from input port #0D to the stack; the most significant byte on the stack is set to zero.

```
CODE IN#0D #0D IN   A L MOV   0 H MVI  H,L PUSH NEXT
```

4.4 Assembler for the 8080; 5. 3. 78 for IPS-N

```
:INT RCODE ENTRYSETUP YES? ! THEN ;
:INT CODE ENTRYSETUP YES? HERE SWAP !
                      THEN ;
: , HERE !B $H INCR ;
16 FIELD VBM " INSTRUCTION ERR! " VBM !T
: FVB YES? 0 VBM SYSWRITE THEN ;
: 7t DUP 7 > SOT <0 OR FVB ;
  #55AA CON IM
: 7txa SWAP 7t 8 * + , ;
: 7tad SWAP DUP IM = YES?     DEL #46 + ,
                       NO: 7t +
                       THEN , ;
: pt RTU DUP <0 SOT 6 > OR SOT OR ( ODD ) FVB
     SWAP DUP <0 SOT 3 > OR FVB RTU 8 * + , ;

: MOV 7t 8 * #40 + 7tad ;
: ADD #80 7tad ;
: ADC #88 7tad ;
: SUB #90 7tad ;
: SBB #98 7tad ;
: ANA #A0 7tad ;
: XRA #A8 7tad ;
: ORA #B0 7tad ;
: CMP #B8 7tad ;
: RET #C0 7txa ;
```

```
: CALL #C4 7txa $DEP ;
: RST #C7 7txa ;
: POP #C1 pt 2 = FVB ;
: SRET #C9 , ;
: SCALL #CD , ;          : SPHL #F9 , ;
: OUT #D3 , , ;          : PCHL #E9 , ;
: IN #DB , , ;
: XTHL #E3 , ;
: XCHG #EB , ;
: DI   #F3 , ;
: EI   #FB , ;
```

```
:   HLT #76 , ;
:   INR #04 7txa ;
:   DCR #05 7txa ;
:   MVI #06 7txa , ;
:   LXI #01 pt 2 > FVB $DEP ;
:   STAX #02 pt >0 FVB ;
:   INX #03 pt 2 > FVB ;
:   DAD #09 pt 2 > FVB ;
:   LDAX #0A pt >0 FVB ;
:   DCX #0B pt 2 > FVB ;
:   PUSH #C5 pt 2 = FVB ;
:   NOP #00 , ;         : RLC #07 , ;
:   RRC #0F , ;         : RAL #17 , ;
:   RAR #1F , ;         : DAA #27 , ;
:   CMA #2F , ;         : STC #37 , ;
:   CMC #3F , ;
:   SHLD #22 , $DEP ;
:   LHLD #2A , $DEP ;
:   STA #32 , $DEP ;
:   LDA #3A , $DEP ;
    0 CON B    0 CON NZ        100 CON NEVER
    1 CON C    1 CON Z       #0354 CON RS
    2 CON D    2 CON NCY
    3 CON E    3 CON CY
    4 CON H    4 CON PO
    5 CON L    5 CON PE
    6 CON M    6 CON POS
    7 CON A    7 CON NEG
:   B,C 0 0 ;       : PC B,C ;
:   D,E 0 2 ;
:   H,L 1 4 ;
:   $SP 2 6 ;
:   PSW 3 6 ;
:   BEGIN HERE ;
:   N: #C3 , HERE SWAP H2INC HERE SWAP ! ;
:   TH HERE SWAP ! ;
```

```
:   JPCODE DUP NEVER = YES?    DEL #C3 ,
                       NO: 1 XOR #C2 7txa
                       THEN ;
:   Y? JPCODE HERE H2INC ;
:   END JPCODE $DEP ;
:   NEXT #0380 NEVER END ;
                  ( END ASSEMBLER 8080 )
```

5 The 6800 assembler

5.1 Resource allocations

With the 6800 access to page #00 has addressing advantages and thus is used as the system page. Except for the stack pointer on the chip, no 6800 registers are permanently occupied and thus may be used without saving operations. The 6800 stack is the parameter stack; the return stack is simulated by a pointer stored on page #00.

Page zero is occupied by IPS between #00 and #9F, you may use locations #A0 to #FF for your code routines as storage area.

Note that with the 6800 16-bit operations use memory in reverse order compared to IPS - the most significant byte is stored at the lower address.
Named variables (pointers) utilizing this reverse storage are identified in IPS by words starting with a reverse slash (\ ). If you wish to use with these words @ and ! , you need to swap the bytes.

Although the stack pointer points to a position one byte lower than the lowest position in use, this is invisible with normal programming; as far as atransfer of this pointer to high-level IPS is concerned, the last position in use is transferred.

Also, the return stack pointer indicates the lowest position in use.

5.2 The instruction mnemonics

The instruction mnemonics either are the original 6800 words or are only slightly modified. They are preceded by an addressing code; and where applicable, this is preceded by an address or an immediate argument.

These are the mnemonics requiring no parameter:

```
TAB   TBA   TAP   TPA   TSX   TXS   PLA   PLB   PUA   PUB
A+B   A-B   INA   INB   INS   INX   DEA   DEB   DES   DEX
ACOM  BCOM  ANEG  BNEG  DAA   SLA   SLB   SRA7  SRB7  SRA
SRB   RLA   RLB   RRA   RRB   CBA   ATST  BTST  RTS   RTI
SWI   WAI   NOP   CLA   CLB   0>C   1>C   0>OF  1>OF  EI   DI
```

The following words need to be preceded by an address or the immediate parameter followed by an address code (C indicates the carry):

```
LDA   LDB   LDS   LDX   STA   STB   STS   STX   +A    +B
+CA   +CB   A-    B-    A-C   B-C   INC   DEC   COM   NEG
ANA   ANB   ORA   ORB   XRA   XRB   SL    SR7   SR    ROL
ROR   ACMP  BCMP  CPX   OST   ABIT  BBIT  JSR   JMP   CLR
```

The addressing modes are identified by one of these preceding address codes:

```
IM PO ABS IX
```

(e.g.: 5 IX LDA load A at the memory position pointed to by the index register + 5)

Control structures are created by:

```
Y? N: TH and BEGIN END .
```

The words Y? and END must be preceded by one of the following condition codes:

LS	lower or same	OF	overflow
C	carry	NEG	negative
Z	zero	LZ	less than zero
LEZ	less or equal zero		

These conditionals are negated, if they are followed by NOT .

5.3 A 6800 assembler example

This code routine divides a number by four by shifting both bytes two positions to the right. Note that the stacks store the most significant byte at the lower address.

```
CODE /4
TSX 0 IX SR   1 IX ROR   0 IX SR   1 IX ROR    NEXT
```

5.4 The 6800 assembler for IPS-N, 27.1.78

```
( ADDRESS-CODES )
#5500 KON IM       #6610 KON PO
#9930 KON ABS      #AA20 KON IX
( BRANCH-CODES ) : NOT 1 + ;
#20 KON NEVER   #22 KON LS   #24 KON C
#26 KON Z
#28 KON OF      #2A KON NEG #2C KON LZ
#2E KON LEZ
( FIELDS )
16 FELD $VB     "   WORD COMB. ERROR      "   $VB   !T
16 FELD $UK     "   ADDR-MODE ERROR!      "   $UK   !T
16 FELD $UB     "   CONDITION ERROR!      "   $UB   !T
16 FELD $ZW     "   BR.OUT OF RANGE!      "   $ZW   !T
( cont. )
```

```
( 6800 assembler, cont. )
( AUX.ROUTINES )
: , HIER !B $H INCR ;
: $AWR SYSLINE 32 + 16 >>> 0 IE ;
:INT CODE ENTRYSETUP JA? HIER VERT ! DANN ;
:INT RCODE ENTRYSETUP JA? ! DANN ;
: $ALTERNATIVDEP
      JA?   + , DUP 256 / , ,
      NEIN: DUP P0 = ZWO IX = ODER ZWO IM = ODER
             JA?   + , ,
             NEIN: WEG $VB $AWR
             DANN
      DANN ;
: $3ALL VERT DUP IM = ZWO ABS = ODER $ALTERNATIVDEP ;
: $2ALL VERT DUP             ABS =    $ALTERNATIVDEP ;
: $0ABIX ZWO IM = JA? WEG $UK $AWR RETEX
                 DANN $2ALL ;
: $ABIX ZWO P0 = JA? VERT WEG IM VERT
                 DANN $0ABIX ;
: $INSERT ( RECEIVES DEPOSIT POSITION AND
                         ON TOP, JUMP-DESTINATION )
   VERT ZWO ZWO 1 + - S>R I 127 > I -128 < ODER
      JA? ZWO ZWO < ZWO 1 - @B #20 = UND
          JA? DUP 1 - #7E VERT !B $H INCR
              S>R DUP 256 / I !B R>S 1 + !B
          NEIN: WEG WEG I $ZW $AWR
          DANN
      NEIN: I VERT !B WEG DANN R>S WEG ;
( STRUCTURING WORDS )
: BEGIN HIER ;
: Y?  DUP #FFD0 UND =0 ZWO #21 =/ UND
                            JA? , HIER $H INCR
                            NEIN: $UB $AWR DANN ;
: N: #20 , HIER VERT $H INCR HIER $INSERT ;
: TH HIER $INSERT ;
: END Y? VERT $INSERT ;
```

```
(   SIMPLE ONE-BYTE   CODE-DEPOSITS    )
:   TAB #16 , ;   :   TBA #17 , ;      :   TAP #06 , ;
:   TPA #07 , ;   :   TSX #30 , ;      :   TXS #35 , ;
:   PLA #32 , ;   :   PLB #33 , ;      :   PUA #36 , ;
:   PUB #37 , ;   :   A+B #1B , ;      :   A-B #10 , ;
:   INA #4C , ;   :   INB #5C , ;      :   INS #31 , ;
:   INX #08 , ;   :   DEA #4A , ;      :   DEB #5A , ;
:   DES #34 , ;   :   DEX #09 , ;      :   ACOM #43 , ;
:   BCOM #53 , ;  :   ANEG #40 , ;     :   BNEG #50 , ;
:   DAA #19 , ;   :   SLA #48 , ;      :   SLB #58 , ;
:   SRA7 #47 , ;  :   SRB7 #57 , ;     :   SRA #44 , ;
:   SRB #54 , ;   :   RLA #49 , ;      :   RLB #59 , ;
:   RRA #46 , ;   :   RRB #56 , ;      :   CBA #11 , ;
:   ATST #4D , ;  :   BTST #5D , ;     :   RTS #39 , ;
:   RTI #3B , ;   :   SWI #3F , ;      :   WAI #3E , ;
:   NOP #02 , ;   :   CLA #4F , ;      :   CLB #5F , ;
:   0>C #0C , ;   :   EI #0E , ;       :   0>OF #0A , ;
:   1>C #0D , ;   :   DI #0F , ;       :   1>OF #0B , ;

(  COMPOSITE CODES )
: LDA #86 $2ALL ;   : LDB #C6 $2ALL ;   : LDS #8E $3ALL ;
: LDX #CE $3ALL ;   : STA #87 $0ABIX ; : STB #C7 $0ABIX ;
: STS #8F $0ABIX ; : STX #CF $0ABIX ; : +A #8B $2ALL ;
: +B #CB $2ALL ;    : +CA #89 $2ALL ;   : +CB #C9 $2ALL ;
: A- #80 $2ALL ;    : B- #C0 $2ALL ;    : A-C #82 $2ALL  ;
: B-C #C2 $2ALL ;   : INC #4C $ABIX ;   : DEC #4A $ABIX ;
: COM #43 $ABIX ;   : NEG #40 $ABIX ;   : ANA #84 $2ALL ;
: ANB #C4 $2ALL ;   : ORA #8A $2ALL ;   : ORB #CA $2ALL ;
: XRA #88 $2ALL ;   : XRB #C8 $2ALL ;   : SL #48 $ABIX  ;
: SR7 #47 $ABIX ;   : SR #44 $ABIX ;    : ROL #49 $ABIX  ;
: ROR #46 $ABIX ;   : ACMP #81 $2ALL ; : BCMP #C1 $2ALL ;
: CPX #8C $3ALL ;   : TST #4D $ABIX ;   : ABIT #85 $2ALL ;
: BBIT #C5 $2ALL ; : JSR #8D $ABIX ;   : JMP #4E $ABIX ;
: CLR #4F $ABIX ;
: NEXT #7E , 0 $DEP ;
                     ( ENDE 6800 ASSEMBLER )
```

6 The 6502 assembler

6.1 Resource allocations

With the 6502 access to page #00 has addressing advantages and this page thus is used as the system page. The return stack is implemented by using the 6502 stack. This stack occupies, by virtue of the 6502 hardware, page #01. The IPS parameter stack is simulated by a pointer (PS) residing on page #00.

Page #00 is occupied by IPS between #00 and #9F, you may use locations #A0 to #FF for your code routines as a storage area.

Both stacks use the lower addresses in memory for the lower bytes of 16-bit numbers. The simulated parameter stack pointer (PS) points to the lowest position in use. The return stack pointer points to the first free position, due to the 6502 hardware design.

All code routines are entered by the emulator with C=0 and Y=1, X is undefined.

With the 6502 the manipulation of 16-bit pointers is somewhat involved. To prevent the code of the IPS primitives from becoming excessively large, six subroutines are available to provide the more common pointer functions required by code routines.

These subroutines represent a compromise between code size and speed; they were designed with the objective of not slowing down the system by more than 10%. Although a more complete set could result in further code reduction, the speed degradation would become significant.

In addition to the word NEXT for leaving a code routine, there are the words DNEXT and PDNEXT. These words return to the emulator, and also increment the parameter stack by 2 (DNEXT) or by 4 (PDNEXT). These words are useful if a code routines leaves fewer parameters on the stack than it receives; it eliminates the explicit pointer manipulation.

6.2 The instruction mnemonics

The instruction mnemonics are largely the original 6502 words; only a few were changed to improve the readability.

These are the mnemonics requiring no preceding parameters:

```
BRK   0>C   0>D   DI    0>OF  DEX   DEY   INX   INY   NOP
PUA   PUS   PLA   PLS   RTI   RTS   1>C   1>D   EI    TAX
TAY   TSX   TXA   TXS   TYA
```

The following words need to be preceded by an address or the immediate parameter followed by an address code:

```
STA   SHL   SHR   ROL   ROR   STX   STY   BCMP  INC   DEC
CPY   CPX   LDX   LDY   A+    ANA   CMP   XRA   LDA   A-
ORA
```

The addressing modes are identified by one of these addressing codes:

A (accumulator without anything else preceding,
e.g. A SHR)

IM PO 0X@ 0@Y 0X 0Y ABS AX AY

Control structures are created by :

Y? N: TH and BEGIN END .

The words Y? and END must be preceded by one of the following condition codes; the N as first letter (for NOT) indicates the opposite:

```
C      carry          NC      not carry
Z      zero           NZ      not zero
NEG    negative       NNEG    not negative
OF     overflow       NOF     not overflow
```

The word @JMP, preceded by an address, assembles an indirect jump using the contents of the specified address as jump destination.

The word JSR (jump subroutine) expects the starting address of a subroutine and assembles a call to that address. As mentioned before, six subroutines are available to simplify the more common 16-bit pointer manipulations.

Page #00 contains a field $AF (8 bytes) for temporary storage of parameters within code routines. Some of the subroutines communicate parameters into this field, because some operations require the addresses residing on page #00.

The subroutine entry addresses are stored in the assembler as constants. By mentioning its name followed by JSR the call is assembled.

1S->$AF transfers one parameter (2 byte)
from the stack into $AF .

It expects Y=1 and delivers Y=0 .

C and the stack remains unchanged.

S->$AF transfers n bytes from the stack into $AF.

It expects Y=n-1 and delivers Y=0. C and stack remain unchanged.

PS-2 decrements the parameter stack pointer by 2 .

It does not expect any particular register setting and only clobbers C.

PS+2 increments the parameter stack pointer by 2 .

It does not expect any particular register setting and only clobbers C.

PC+2 increments the pseudo-PC by 2.

It does not expect any particular register setting and only clobbers C.

D-PREP expects Y=1 .

It transfers the stack pointer into $AF and then increments the stackpointer by 2. It then sets Y=0 and loads A by performing a PS 0@Y LDA .
It returns with this A , C=0 and Y=0 .

This subroutine is most useful for operations working on two parameters on the stack and leaving a single result.

6.3 A 6502 assembler example

This routine swaps the two bytes of a number on the stack.

```
CODE BSWAP
     PS  0@Y LDA TAX DEY  PS 0@Y LDA INY PS 0@Y STA
     DEY TXA 0@Y STA NEXT
```

6.4 The 6502 Assembler for IPS-N, 23. 1. 78

```
(    ADDRESS-CODES )
0    KON A       1 KON IM              2 KON P0
3    KON 0X@     4 KON 0@Y             5 KON 0X
6    KON 0Y      7 KON ABS             8 KON AX
9    KON AY

( BRANCH-CONDITIONALS )                #4C KON NEVER
#90 KON C     #B0 KON NC               #F0 KON NZ     #30 KON NNEG
#D0 KON Z     #10 KON NEG              #50 KON OF     #70 KON NOF
( FIELDS     )
210 FELD     $CTAB
16 FELD      $VB     "   PREC.WORD EERROR   "   $VB    !T
16 FELD      $UK     "   ADDR-MODE ERROR!   "   $UK    !T
16 FELD      $UB     "   CONDITION ERROR!   "   $UB    !T
16 FELD      $ZW     "   BR. OUT OF RANGE   "   $ZW    !T
( AUX-ROUTINES )
: $AWR SYSLINE 32 + 16 >>> 0 IE ;
: , HIER !B $H INCR ;
:INT CODE ENTRYSETUP JA? HIER VERT !
                       DANN ;
:INT RCODE ENTRYSETUP JA? ! DANN ;
: !FB (FIELD BYTES) DUP S>R + 1 R>S
                     JE 1 - DUP S>R !B R>S NUN WEG ;
: $AINS      ZWO 1 - @B #4C =          ( DEPOSIT POSITION AND
```

```
                     ON TOP DESTINATION )
JA?    VERT !
NEIN: ZWO 1 + - DUP 127 > ZWO -128 < ODER
                JA?   $ZW $AWR VERT WEG
                NEIN: VERT !B
                DANN
DANN ;
```

```
: $MDEP VERT S>R I 9 > I <0 ODER
             JA?   $VB $AWR
             NEIN: I + $CTAB + @B DUP =0
                   JA?   $UK $AWR
                   NEIN: , I >0 JA? I 6 > JA?   $DEP
                                             NEIN: ,
                                             DANN
                                   DANN
                   DANN
             DANN R>S WEG ;
( STRUCTURING WORDS )
: @JMP #6C , $DEP ; : JSR #20 , $DEP ; : BEGIN HIER ;
: Y? DUP #4C =
       JA?    , HIER H2INC
       NEIN: DUP #10 - #FF1F UND =0
                  JA? , HIER $H INCR
                  NEIN: $UB $AWR DANN DANN ;
: N: #4C , HIER VERT H2INC HIER $AINS ;
: TH HIER $AINS ;
: END Y? VERT $AINS ;
(   SIMPLE CODE-DEPOSITS )
:   BRK #00 , ;  : 0>C #18 , ;   :   0>D   #D8   ,   ;
:   DI #58 , ;   : 0>OF #B8 , ;  :   DEX   #CA   ,   ;
:   DEY #88 , ;  : INX #E8 , ;   :   INY   #C8   ,   ;
:   NOP #EA , ;  : PUA #48 , ;   :   PUS   #08   ,   ;
:   PLA #68 , ;  : PLS #28 , ;   :   RTI   #40   ,   ;
:   RTS #60 , ;  : 1>C #38 , ;   :   1>D   #F8   ,   ;
:   EI #78 , ;   : TAX #AA , ;   :   TAY   #A8   ,   ;
:   TSX #BA , ;  : TXA #8A , ;   :   TXS   #9A   ,   ;
:   TYA #98 , ;

(   ZUSAMMENGESETZTE BEFEHLE )
: STA  0 $MDEP ; : SHL  10  $MDEP ; : SHR 20 $MDEP   ;
: ROL 30 $MDEP ; : ROR  40  $MDEP ; : STX 50 $MDEP   ;
: STY 60 $MDEP ; : BCMP 70  $MDEP ; : INC 80 $MDEP   ;
: DEC 90 $MDEP ; : CPY  100 $MDEP ; : CPX 110 $MDEP  ;
: LDX 120 $MDEP ; : LDY 130 $MDEP ; : A+  140 $MDEP  ;
```

```
: ANA 150 $MDEP ; : CMP 160 $MDEP ; : XRA 170 $MDEP  ;
: LDA 180 $MDEP ; : A-  190 $MDEP ; : ORA 200 $MDEP  ;
( cont. )
```

(6502 assembler, cont.)

```
( TABLE $CTAB , ADDRESSING SEQUENCE :

 INSTR A    IM  P0  0X@ 0@Y 0X  0Y  ABS  AX   AY)
( STA ) 0   0   #85 #81 #91 #95 0   #8D  #9D  #99
( SHL ) #0A 0   #06 0   0   #16 0   #0E  #1E  0
( SHR ) #4A 0   #46 0   0   #56 0   #4E  #5E  0
( ROL ) #2A 0   #26 0   0   #36 0   #2E  #3E  0
( ROR ) #6A 0   #66 0   0   #76 0   #6E  #7E  0
( STX ) 0   0   #86 0   0   0   #96 #8E  0    0
( STY ) 0   0   #84 0   0   #94 0   #8C  0    0
( BCMP) 0   0   #24 0   0   0   0   #2C  0    0
( INC ) 0   0   #E6 0   0   #F6 0   #EE  #FE  0
( DEC ) 0   0   #C6 0   0   #D6 0   #CE  #DE  0
( CPY ) 0   #C0 #C4 0   0   0   0   #CC  0    0
( CPX ) 0   #E0 #E4 0   0   0   0   #EC  0    0
( LDX ) 0   #A2 #A6 0   0   0   #B6 #AE  0    #BE
( LDY ) 0   #A0 #A4 0   0   #B4 0       #AC  #BC  0
( A+ ) 0    #69 #65 #61 #71 #75 0       #6D  #70  #79
( ANA ) 0   #29 #25 #21 #31 #35 0       #2D  #3D  #39
( CMP ) 0   #C9 #C5 #C1 #D1 #D5 0       #CD  #DD  #D9
( XRA ) 0   #49 #45 #41 #51 #55 0       #4D  #5D  #59
( LDA ) 0   #A9 #A5 #A1 #B1 #B5 0       #AD  #BD  #B9
( A- ) 0    #E9 #E5 #E1 #F1 #F5 0       #ED  #FD  #F9
( ORA ) 0   #09 #05 #01 #11 #15 0       #0D  #1D  #19

$CTAB 210 !FB

( AUX-OPERATIONS, CONSTANTS AND SUBROUTINES )
: NEXT   #0080 NEVER END ; : JNEXT #0000 NEVER END ;
: WNEXT #0000 NEVER END ; : PWNEXT #0000 NEVER END ;
#0000 KON HP        #0000 KON PS
#0000 KON PC        #0000 KON $AF
#0000 KON 1S->$AF   #0000 KON S->$AF   #0000 KON PS-2
#0000 KON PS+2      #0000 KON PC+2     #0000 KON D-PREP
               ( END ASSEMBLER FOR IPS-N )
```

7 The CDP 1802 assembler

The RCA COSMAC CDP 1802 is used in the flight computer of all AMSAT Phase III communications satellites.

The 1802 is similar to the 1801, with an extended instruction set. Refer to section 3 (CDP 1800) for details of resource allocations and instruction mnemonics.

7.1 Listing of the CDP 1802 Assembler

```
: ,     HIER !B $H INCR ;
:INT    RCODE ENTRYSETUP JA? ! DANN ;
:INT    CODE ENTRYSETUP JA? HIER VERT !
                        DANN ;

( ERROR MESSAGES )
16 FELD $JERROR " DESTIN. OFF PAGE " $JERROR !T
16 FELD $CERROR " PREINSTR. ERROR! " $CERROR !T

( AUX ROUTINES )
: $TERM WEG $CERROR SYSWRITE ;
: $INADR PDUP EXO #FF00 UND
        =0 JA?    !B
           NEIN: WEG $JERROR SYSWRITE
           DANN ;
: $AD           ZWO #FFF0 UND =0 JA?    + ,
                                NEIN: $TERM
                                DANN ;
: $ALUACT ZWO #FF00 UND #A500 =
           JA?    VERT $AD
           NEIN: $TERM
           DANN ;
( cont. )
```

```
( CDP 1802 assembler, cont. )

: MX        0 $ALUACT ;
: IM        8 $ALUACT , ;

: LDN  #00  $AD ;    : RET  #70  , ;  #A574  KON   ADC
: INC  #10  $AD ;    : DIS  #71  , ;  #A575  KON   -DC
: DEC  #20  $AD ;    : LDXA #72  , ;  #A577  KON   DC-
: DA   #40  $AD ;    : STXD #73  , ;  #A5F0  KON   LD
: STR  #50  $AD ;    : ROR  #76  , ;  #A5F1  KON   OR
: I/O  #60  $AD ;    : SAV  #78  , ;  #A5F2  KON   AND
: GLO  #80  $AD ;    : MARK #79  , ;  #A5F3  KON   XOR
: GHI  #90  $AD ;    : 0>Q  #7A  , ;  #A5F4  KON   ADD
: PLO  #A0  $AD ;    : 1>Q  #7B  , ;  #A5F5  KON   -D
: PHI  #B0  $AD ;    : ROL  #7E  , ;  #A5F7  KON   D-
: ->P  #D0  $AD ;    : NOP  #C4  , ;  #0002  KON   RS
: ->X  #E0  $AD ;    : SKP2 #C8  , ;  #0000  KON NEVER
: LSIE #CC    , ;                     #0009  KON   Q=1
: IDL  #00    , ;
: SHR  #F6    , ;                     #000A  KON   D=0
: SKP  #38    , ;
: SHL  #FE    , ;                     #000B  KON   DF
: INCX #60    , ;
: NEXT #D6    , ;                     #000B  KON   PS
: NOT     8 EXO ;
: EF 11 + ;
: BEGIN  HIER  ;
: Y?     #30 $AD HIER $H INCR ;
: TH     HIER VERT $INADR ;
: N:     0 Y? VERT TH ;
: END    Y? $INADR ;
: $SBRT ZWO 4 UND >0 JA? VERT ( PROVOZIERT FEHLER )
                     DANN $AD ;
: LEND #C0 $SBRT DUP 256 / , , ;
: LSKP #C4 $SBRT ;
                    ( END CDP 1802 ASSEMBLER )
```

CHAPTER III

The IPS System

1 General design considerations

IPS was primarily designed to allow the speedy writing of programs intended for the control of satellites, scientific data collection and other engineering applications. There are many programming languages claiming to be suitable for these applications. But on closer inspection most of these require either rather large systems and thus are not very practical for microcomputers or they have serious limitations, like insufficient speed or no multiprogramming.

Most control oriented languages are derived from languages created for mathematical or commercial data processing. Generally, this means that the real-time part needs to be handled by the operating system, and the power of this combination is highly dependent on the capabilities of the operating system.

With IPS a different approach was possible, since there is no real need to maintain compatibility with other languages and an entirely different approach could be taken.

Every programming language represents an interface between machine and men. Thus, it must comply with two requirements:

1. The language should allow the translation of programs making efficient use of the underlying processor, both from a speed and memory economy point of view. This is essentially an engineering problem.

2. The language should allow the expression of problems in a way matching the human understanding and decomposition of problems; the system is to be "user-friendly". Achieving this

is not an engineering problem, but a problem of psychology and aesthetics - a form of art.

Let us look at the second point first.
In order to be able to put the problem into perspective it would be necessary to define the "human way of understanding"; obviously an impossible task because it would have to take into account the different backgrounds of all the people intending to use the system.

The second best approach would be to isolate certain general aspects of a problem area and make sure that these are matched by the language.

Along this track lots

of subjective judgements are bound to creep into the language. This is not as bad as it implies, though.

We adapt easily to a given structure and thus soon feel comfortable with it. This adaptation, on the other hand, causes us to cling to given structures we are used to. This makes progress in this discipline very slow because once somebody is adapted to a language lots of energy is necessary to learn or invent some new way.

Most programming languages initially were designed to allow mathematical problem solving. There we have at the core of the problem the requirement to solve algebraic expressions as given by mathematical formulas. These describe a desired result in terms of the formulas. If the syntax of these formulas is unambigious a program may turn this syntax into actions (procedures) producing the result.

This general approach resulted in that most programming languages today are essentially syntax driven and the main thrust of investigation went into the understanding of syntax.

Although significant progress has been made in syntax driven high-level languages, their generally somewhat static frame has prevented them from being used extensively in systems programming and real-time engineering. In these disciplines the tremendous power of the strictly procedural assembly language was needed and outweighed the need to live with the sometimes quite baroque instruction sets of the hardware.

In this problem area the programmer, in planning and thinking, wants the computer to perform certain actions. If he uses a high level language, he has to transform these procedures into the syntax of the programming language. The compiler in turn decomposes this syntax again into a set of instructions.

Thus, it becomes clear that two translation processes take place. A short cut is possible if the detour via the syntax can be avoided. With IPS it was attempted to design a language that is strictly procedural and thus offers the power of assembly language.

But the underlying instruction set was designed to be more acceptable from a human point of view than the instruction sets defined by hardware constraints.

One of the most powerful procedural computer architectures can be built around the concept of the stack. This technique has been in use for a long time in systems programming, and it became widely known by its use in the Hewlett Packard calculators.

A very powerful and comfortable instruction set can be built around this concept, but it requires some relearning to become comfortable with reverse polish notation (RPN). The acceptance of this approach for mathematical problem solving is a good demonstration of its power.

With engineering problems it is even more appropriate, because its only disadvantage, the reduced readability of algebraic expressions, is less serious.

With these problems the intelligence of the programs resides in the flow of control.

Expressions play a minor role.

Sect. 1 - General design considerations 65

Designing a language for small systems is constrained by the amount of memory that can be allocated for the system. With IPS the target system was to use 16 Kbytes of which a not to large fraction could be allocated to the compiler.

This size necessarily limits the system to the more basic capabilities needed in most programs. More esoteric features are not made part of the language; rather the language was kept open ended to allow special extensions.

This open endedness results from the inclusion of an assembler and from its procedurality:

There is no distinction between
the language words,
the application words and
the command words (if an interpretive mode is included).

Thus, the language provides the means to change itself (it is reflexive).

The interpretive mode is most useful for testing; it simplifies the communication with the system and allows a fast localisation of problem areas and also allows for the unique macro capabilities of the assembler.

The "friendliness" of a language is largely governed by the availability of an interpretive mode.
It allows the resolution of problems in a dialogue. This implies that the language is as unrestrictive as possible, but has sufficient diagnostics to catch most trivial errors caused by the human inability to be 100 % reliable.

With the diagnostic mechanism of most programming languages three areas are relevant.

First, they strictly enforce their syntax at the compilation time. This also generally assures the integrity of the translation systems.
(A syntax error does not bomb the compiler.)

Second, with data, syntax checking alone is not sufficient, so often it is augmented by run time checks.

Thirdly, the semantics of the program should be checked. Even with simple programs it is impossible for practical reasons to check a program for correctness.
So at best plausibility checks can be applied.

Given the limited resources of microcomputers, with IPS a somewhat relaxed attitude was taken. IPS has very little syntax to enforce; it will find only the most trivial error like misspelled words.

Some additional diagnostic was added to find the more typical errors. Some simple plausibility checks are possible because most semantic errors are accompanied by unaccounted for items on the stack.
Although this technique does not provide a tight net, most typical errors get caught this way. On the other hand, this approach does not restrict the language because these checks always can be circumvented.

As to the data types, IPS was designed to be completely transparent with regard to the underlying hardware. Because all relevant processors use a 8/16 bit memory organisation, IPS uses these types as the data primitives.

This allows a very effective use of the hardware and results in a similar power as assemblers in defining other arbitrary data types.

This power is augmented by the fact that memory addresses also have a 16 bit structure and thus can be manipulated just like other numerical data.

Because IPS is primarily aiming at real time

applications, no run-time checking is provided. This would provide considerable overhead, but could not ensure programs to live up to their expectation in case of an error. If there is an error, an abort with a real-time application is as bad as a commissive error. So, it would be hard to justify the overhead.

The use of floating point arithmetic is also a sort of generalized run time checking; if the programmer thinks about the scaling of his numerical types while writing the program, the run-time scaling operations of floating point operations with their overhead are not necessary.

If run-time efficiency is a primary consideration, it makes sense not to use floating point arithmetic. This is particularly true, since the careless use of floating point numbers may occasionally result in an exponent overflow; so after all the programmer is not entirely relieved of the burden of thinking about the ranges of his data.

Although a procedural language design like IPS cannot provide the degree of "protection" like languages with a fixed syntax, it turns out that the plausibility checks provided are adequate for all practical situations.

The use of a stack architecture eliminates a problem area plaguing syntax driven languages. There the passing of parameters between modules, the initialization of variables and their range of validity always requires special consideration.

With a stack computer, local variables are unnamed and kept on the stack, only global variables are named. On the stack the exchange of parameters becomes completely transparent; no formal techniques are required.

One of the major breakthroughs with programming languages occurred with the introduction of the concept of structured programming and top-down design.

Structured programming implies that the arbitrary flow of control in programs, as exemplified by GOTOs and labels, is eliminated.
The control structures are rather designed as to have only a single entry and exit point. Programs written along these lines are naturally modularized and thus much easier to manage mentally.

Most languages supporting structured programming additionally allow labels and GOTOs. With IPS the implementation was significantly simplified by the complete elimination of labels. Initially this caused some mixed feelings.

Using hindsight, though, this step was completely justified; after some practice the absence of labels is not noticed anymore.
The top-down design of programs implies an effective parameter passing and nesting mechanism of subroutines.

A stack architecure like IPS is probably the most effective structure known today towards this end. In fact, the effectiveness results from the presence of a second stack for the internal return address management.
The parameter stack thus may be used exclusively for
subroutine parameter passing and thus is entirely transparent to the programmer.

At compile time the parameter stack is used for the address management of the control statements. With strictly structured programming the jump addresses require a last-in first-out structure; this is equivalent to maintaining them on the stack. Thus, there is a close relationship between structured programming and a stack computer.

Because of the match of structured programming to the human way of thinking one may speculate that the stack computer as a generalisation of this principle also matches well the human way in other regards, e.g. the fact that definitions using the stack are re-entrant and recursive or that syntax with IPS is largely replaced by an interaction mechanism between operators and the stack.

At any rate, most people exposed to IPS become very comfortable with it after a short while.

2 The IPS implementation

For the implementation of a language three different techniques are available:

The classical interpreter stores the program source text. At execution time this text is analysed and the appropriate actions are executed.
The analysing results in considerable overhead; such interpreters thus are typically very slow.

The other common approach is the compiler.
With this technique the program is analysed and translated into machine instructions; at the execution time these instructions can be executed without any additional overhead.

The problem with this approach is the fact, that the produced code usually is "bulky" and contains lots of repetitions of basic instruction sequences.
A good compiler, though, can produce code that may be very close to optimum. Smaller compilers with a size fit for microcomputers usually produce sub-optimum code invoking size and speed penalties of a factor of two or more.

A third approach chosen for IPS tries to combine the best out of both techniques. It uses a set of high level instructions defined by optimized code routines.

A compiler produces pseudocode treating these code routines as the "instruction set" of a virtual high level machine. Because these "instructions" (usually addresses to code) take only very little overhead to interpret, such a system only is two or three times slower than optimum assembler code.

Provided that this is acceptable, this technique results in extremely compact programs requiring considerably less memory than the other approaches.

Furthermore, the pseudocode is machine independent; only some 40 primitive code routines represent an interface to the underlying processor. This makes it very easy to transcribe the system to a new microcomputer; only about 1 Kbyte of code has to be rewritten in each case.

This effectively decouples the software from the volatility of changing hardware technology.

Most of the system, of course, is written in IPS and thus is identical with each processor.

3 Structure of the pseudocode

The pseudocode should produce as little overhead as possible to interpret.
Rather than using a fixed "pseudo instruction set" the overhead may be reduced, if addresses to executable code are used. In this case the "instruction" is an address pointing to executable code; no lookup is necessary.

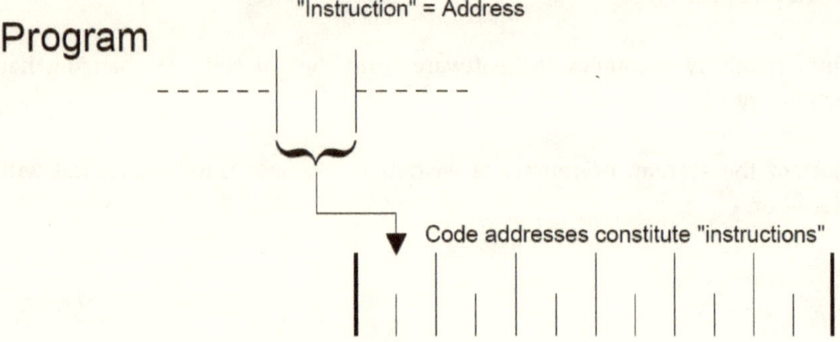

Addresses as pseudocode

This technique can be extended by using the pseudocode not as addresses to executable code, but to a pointer which in turn contains the address of the executable code. This indirect execution requires at the head of each instruction a "descriptor" or typecode containing the pointer to executable code.

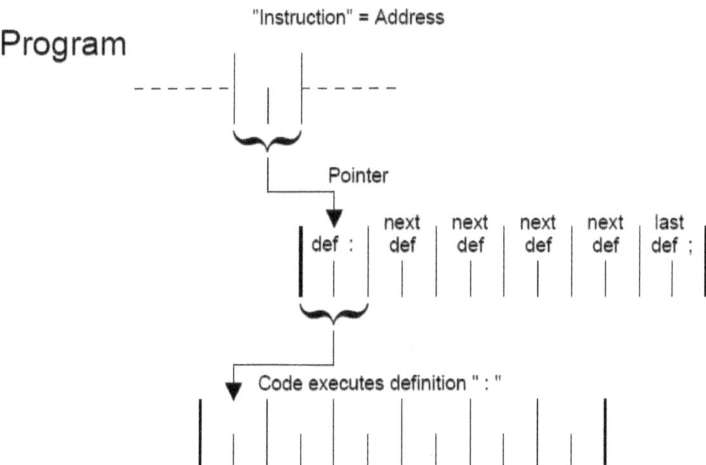

Indirect execution of IPS

This indirect interpretation results in two significant advantages over the direct concept.
If the header of a routine identifies it as a definition (a subroutine), the pseudocode does not need to contain the call address; it is implicitly available; it is header + 2. Thus memory is saved. Even more important is the fact that indirect execution results in the same format for all instructions.

Thus, at compile time the compiler need not check the nature of the instruction in order to be able to produce code. All instructions compile along the same rule.

By putting this semantic descriptor of the instruction where it logically belongs, namely at the head of it, a concise instruction format is achieved without creating additional overhead. In fact, because of this unified instruction format, no distinction between

IPS words,
application words,
code routines or
command words

is necessary. It is the key to the natural extensibility of the language.

As an example, the following diagram shows the code structure of the word H2INC . This definition is used by the compiler to increment the contents of $H by two. It is defined as:

`: H2INC HERE 2 + $H ! ;`   with HERE previously defined as : HERE $H @ ;

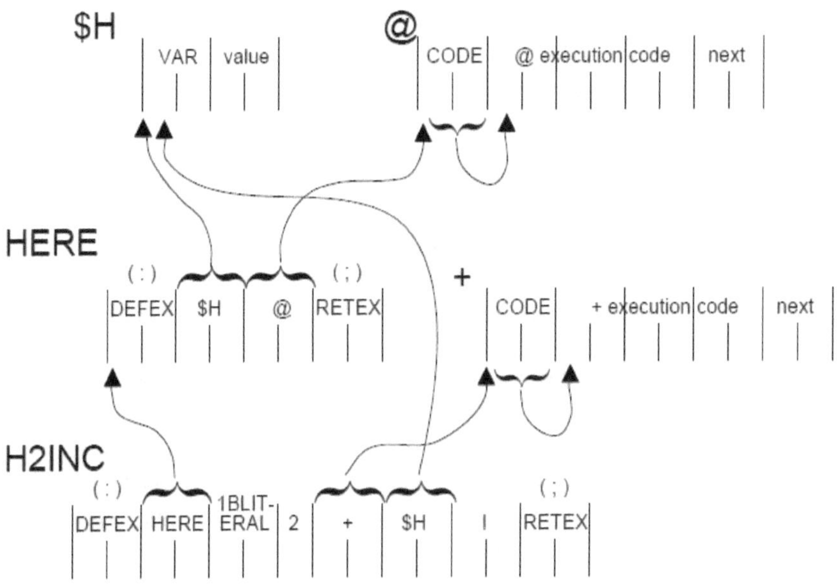

Code Structure of H2INC

4 The Executive

The pseudocode is executed by the inner interpreter. This routine employs a pointer, the so-called pseudo-program counter. This pseudo-PC points to the next pseudo instruction to be executed in turn. Pseudo-PC

The inner interpreter fetches the contents of the location the pseudo-PC points to (16 bit) and then increments the pseudo-PC by two. This fetched number is called header pointer (HP) and it points, as the name implies, to the header of the routine to be executed.

The interpreter now fetches the contents of this location. The fetched number is the address where the executable code of the particular routine starts;
the interpreter jumps to this address and the program continues by executing the instructions at this address.

At the end of this code there is a jump back to the inner interpreter, which then fetches the next pseudo instruction in turn.

It is clear from this description, that the computer spends a significant fraction of its time in this inner interpretation loop. Thus, care was taken to code this inner interpreter to be as fast as possible.

The inner interpreter stores HP at some location, because some instructions need this pointer. With a definition it defines the address where to continue, witha variable it defines the parameter field.

Ideally HP also gets incremented by two by the interpreter. But with some processors this is not possible without overhead and thus is not done. The code routines then have to increment it as required.

With the COSMAC, HP is incremented by two; with the 8080 by one and with the 6800 and 6502 not at all.

Every 20 ms the normal interpretation loop is suspended (after completion of the last code routine) in order to do some housekeeping jobs.

Within this "pseudo-interrupt" the clock and the stopwatches are updated, keyboard entries are treated, and the compiler is informed if there is an input to be processed. If there is special hardware I/O, the 20 ms routine handles these devices. Finally, the tape cassette write output and motor control is updated every 20 ms.

An external hardware input has to inform the inner interpreter if 20 ms have expired. With the COSMAC this is done in a straightforward way using the "external flag 1" input.

The inner interpreter requires a jump back into the interpretation loop; by making this jump conditional on 1 EF NOT, the interpreter leaves the loop if EF1 is set.

Within the 20 ms routine the 20 ms input flip-flop is reset waiting to be triggered with the next 20 ms pulse again.

With the other processors, unfortunately, such a simple technique is not possible. The only way would be to add a few instructions in the interpreter to scan the 20 ms input after each interpretation cycle.

This would considerably slow down the system; a serious penalty considering the fact that the computer already spends about half of its life in the interpretation loop.

Since the interpretation speed is a primary consideration, the only way out is to use a real interrupt every 20 ms to replace the first instruction of the interpreter by a call or jump into the 20 ms routine.

On the "next time around" thus the computer jumps into the 20 ms program. This, among its other duties, restores the interpreter code again to its original state. After the 20 ms routine the normal interpretation is resumed.

The described technique of self-modifying code basically is very undesirable for a number of reasons. But because this technique increases the speed of the computer effectively by some 20%, and because it is localized to a single instance within a short interpretation routine, it is a well considered technical compromise to circumvent the shortcomings of some processors.

In addition to the interpreter and the 20 ms routine, an interrupt routine is provided to process cassette inputs. This handler is very simple; it deposits an input byte at a pointer location and then increments the pointer.

One of two pointers is used, depending if LOADFLAG is set or not. Each pointer has a limit (LOADLIMIT and $PE); the 20 ms routine checks if the input has progressed beyond these pointers and then initiates the appropriate actions.

5 The machine independent part

During the following discussion please refer to the listing of your processor at the end of this chapter (Sect. 7). The listed program describes your compiler in a way usable for an automated translation. There we run into the problem of having to define a language in its own terms. In principle this is possible if the translation process is carefully sequenced and the prohibition to redefine words is disabled.

The first IPS versions were translated this way using a previous version to produce the new version. Unfortunately, this technique becomes very clumsy if the compiler is to be translated for a different processor.

A much more general solution to this problem results if the language is not defined in its own terms, but in a different language; a so-called meta language. This meta language of course may be exactly the same as to the meaning of the words, only different names are to be used to distinguish between the translation operation and the object to be translated.

To aid portability of IPS, such a meta language and the compiler to go with it (IPS-X) was developed. This was a very worthwhile investment; it enables a very fast transcription of IPS to an arbitrary processor.

The IPS-X meta language was derived from the German version of IPS simply by using lower case letters instead of the usual upper case letters.
Where this was not applicable (e.g. with the colon), the meta language words were produced by appending a lower case n .

This list should allow you to read the listings, it shows only the words differing from the English version:

```
weg DEL     ,  hier HERE      ,  aufnahme RECORD ,
!fk !FC     ,  ja? YES?       ,  nein: NO:  ,
dann THEN  ,  je EACH        ,  nun NOW  ,
+nun +NOW  ,  anfang LBEGIN ,  ende? LEND  ,
dann/nochmal  THEN/REPEAT    ,
kon CON     ,  feld FIELD     , ~ " , zwo SOT  .
```

5.1 Compiler and the chain

The main program executed continuously by IPS is the chain. This program consists of 9 instructions (pseudo instructions, of course) and a jump back to the first instruction. It is usually located on the system page.

Initially, the last 7 instructions are no-ops, an instruction not changing anything. The first position contains the address of the COMPILER , the program that processes keyboard and cassette inputs.

The second position (CHAIN position 0) contains the STACKDISPLAY . Compiler description

Please try to find the COMPILER in your listing (next page and also Sect. 7) for the following discussion.

The compiler is executed periodically and processes one input word with each pass through it, provided an input is available. It does so by calling $NAME .
This definition scans the input and encodes the word in the standard IPS format (see further down) and returns a 1 if a word is available. Otherwise it returns a 0 on the stack.

If after the word the end of the input buffer has been reached, $NAME (or $CSCAN) additionally sets the variable $P2 to 1, else it clears it.

If a word is available, $SUCH tries to find it in the nametable. If it finds it, it returns the starting address of the particular routine.
If the word does not exist in the nametable, a 0 is returned.
On top of this address the position of the entry in the nametable is returned.
This is put into $V1 for possible later use.

Now the compiler checks if the system is in the limited input mode.
If so, the word RUMPELSTILZCHEN can reset this mode. If the address found by $SUCH is lower than LIS or the system expects a number input, the necessary actions are taken.

An input error is treated by calling IE . This word marks the input with a question mark and terminates further processing. IE removes one item from the stack.

Provided that so far everything went well, the return of $SUCH is checked for 0. If a 0 has been returned, the input may be a number ("number processor").
In this case $ZAHL is called to convert the input word into a number on the stack. If this conversion was successful, a 1 is returned on top of this number by $ZAHL , else a 0.

Let us assume a number is available. In this case the number is simply left on the stack in the interpretive mode.

But in compile mode (COMPILEFLAG = 1) the compiler has to deposit a literal code and then the number.
If the number is between 0 and 255, the address of 1BLITERAL is deposited followed by a single byte containing the number.
Else the address of 2BLITERAL is deposited followed by the number stored into two bytes.

If the system is in the limited input mode, the number is placed into N-INPUT and N-READ is reset.

If the input word could not be converted into a valid number, again IE is called.

If $SUCH returned an address (i.e. the name is available in the nametable), the "found processor" is invoked.

The action to be taken depends on the type of the entry and if the system is in compile mode or not. There are four different types of entries:

Compiler action matrix

	COMPILEFLAG	=0	=1
Code	Entrytype	Interpretive Mode	Compile Mode
#00	: (normal)	execute	compile
#80	:PRIOR	IE	execute
#40	:HPRI	execute	execute
#C0	:INT	execute	IE

IE = Input Error

Names of ":" definitions and names of VAR , CON and FIELD entries are entrytype normal.

Words like YES? and THEN are entrytype :PRIOR in order to allow stack manipulations at compile time.

The words ' " ICHN DCHN are of entrytype :HPRI . This allows the compiler to take the appropriate actions in both modes.

Words like : CON and VAR are defined as entrytype :INT to prevent entries within entries; i.e. they can only be interpreted.

The "found processor" first fetches the entrytype code out of the nametable.
It is stored as the two most significant bits of the first byte of the namecode. The position of this code was stored after $SUCH in $V1 .

This code, in conjunction with the COMPILEFLAG , defines the position in the action matrix. If the word is to be compiled, at first there is a check of HERE against $ML (memory limit) to see if there is still room left. Then the address is deposited by $DEP
.

If the word is to be executed, this is done by the word $TUE . This word expects the typecode field address and executes the word. Because this execution may affect the return stack, it is turned into a virgin state by storing the return address of the compiler into $V1 .

While the compiler is being executed, its return address to the chain is on the return stack. After execution the return stack is restored from $V1 .

After these actions the stack pointer is checked against $SL (stack limit).
If the pointer exceeds the limit, the pointer is restored and an error message is written.
The first underflow position of the stack is preset to $F1 for a reason explained later.

Compiler activities are governed by the state of the READYFLAG . If it is 1, an input is available for processing. IE resets this flag. If $P2 has become 1 ,
$NAME reached the end of the input buffer; COMPILER has to reset READYFLAG in this case and also clear the input buffer to indicate the successful completion of input processing.

A generic compiler program listing is as follows. See also Sect. 7.

```
: COMPILER $NAME
YES? $SUCH
   1     ( TO CONTINUE ) LIM @B
        YES? SOT ' RUMPELSTILZCHEN
                  = YES?     ( RUMP. )    0 LIM !
                    NO: ( NOT RUMP. ) N-READ @
                          YES?    PDEL 0 1
                          NO: SOT LIS @ <
                                 YES? IE DEL 0
                                 THEN
                          THEN
                    THEN
        THEN
  YES? ( CONTINUE ? )            DUP =0
       YES? ( NUMBER PROCESSOR )
         DEL $ZAHL
         YES? COMPILEFLAG @B
            YES? DUP #FF00 AND
             =0 YES? ' 1BLITERAL $DEP
                       HERE !B $H INCR
                 NO: ' 2BLITERAL $DEP $DEP
                 THEN
            NO: LIM @B YES? N-INPUT ! 0 N-READ !
                           THEN
            THEN
         NO: IE
         THEN
       NO: ( FOUND PROCESSOR ) DUP 6 - @B #C0 AND
         COMPILEFLAG @B OR DUP 1 =
         YES? DEL HERE $ML >=
                          YES? DEL MEMMESSAGE SYSWRITE
                          NO: $DEP
                          THEN
         NO: DUP #80 = SWAP #C1 = OR
                 YES?     IE
                 NO: R>S $V1 ! $TUE $V1 @ S>R
```

```
                THEN
            THEN
         THEN
         $PSHOLEN $SL > YES? $SL $PSSETZEN
                            STACKMESSAGE SYSWRITE DEL $F1
                            THEN
  THEN
THEN
READYFLAG @B $P2 @B AND ( $BU ALL PROCESSED? )
  YES?
     #20 $BU !B $BU DUP 1 + 511 L>>> $CEN
  THEN
;
```

5.2 Compiler support routines

5.2.1 Scanner and namecoding

The processing of the input string in the input buffer (second half of the TV-screen) is effected by the definition $NAME .

Two operations are necessary:

1. The input string needs to be segmented to isolate individual words.

2. The thus isolated word must be transformed into the 4-byte IPS namecode to simplify further activities. $NAME listing

```
: $NAME           0       READYFLAG @B
     YES? 1 $CSCAN >0
        YES? $PI @ $P1 !
          2 $CSCAN PDEL #CE57 #8D
          $P1 @ $PI @ SOT - DUP 63 > YES? DEL 63
                                      THEN
          DUP $ND !B 1 - SOT +
          EACH I @B $POLYNAME
          NOW $ND 3 + !B $ND 1 + ! 1
        THEN
         THEN ;
```

The scanning process uses two pointers.

The input characters are first scanned until the first non-separating character is found.
(Separators are blanks, comments and end of input block).

The pointer variable $P1 is left pointing to this first valid character. Scanning then continues for valid characters until a separating character is again found.

The pointer variable $PI then is left pointing at the first position after the last valid character of the word.

The actual scanning is effected by $CSCAN . This definition takes a 1 from the stack to scan over the separators and terminates scanning upon finding the first non-separator. It returns a number >0 if it found a valid character; else a 0.

$CSCAN additionally sets $P2 to 1 if it reached the end of the input buffer. The last character of the input buffer is pointed to by the contents of $PE .

The definition $CSCAN always uses $PI as the scan pointer. Because the first character is to be identified by $P1 , the content of $PI is put into $P1 after scanning the separators.

Then $CSCAN again is called with a 2 on the stack. This signals to it to scan now all valid characters until a separator is found again. It again returns a number; that is discarded. Now the input word is bracketed by $P1 and $PI as desired.
The next operation for $NAME is to encode the identified word in the format used by all names stored in the nametable.

To understand this process, some general remarks are in order.

Names are used to identify objects and may use up to 63 characters. If the names were stored in the original form, inordinate amounts of memory would be required.
Thus some encoding of the name is desirable to reduce the amount of storage required.

Conventional languages generally accomplish this by storing the first few characters of the name, usually 6 or 8, as a namecode and ignore the rest.

With IPS this technique is not practical. The needed storage space still would be rather large. More important, the inadvertent omission of a separating blank could lead to a concatenation of words without error message. This would constitute very hard to find errors.

The technique used with IPS attacks the problem along an entirely different line. All possible names may be viewed as a vector space with 25663 different points. The permutations possible with 63 bytes is a number large beyond grasp; about 10150.

On the other hand, a practical system will hardly ever use more than 500 names. After all these names have to be learned and managed by humans. This means that the name space is practically empty in comparison to the total number of possibilities.

The idea of the IPS name coding is to map the extremely large original name space in a pseudorandom fashion into another space. If this new "code space" is still very large as compared to say 500, this space again will be nearly empty.
Thus, the probability of two different names mapping to the same point in the code space can be extremely small.

If 4 bytes are selected as the size of the namecode, the resultant code space contains 2564 (about 4 x 109). With 500 points occupied, the probability of a collision is on the order of 10-7 and thus may be neglected for all practical purposes. Still, if a collision would occur, a very confusing situation could arise.

With the stack language IPS this problem does not exist though, because with IPS all named objects are global and the redefinition of names would not make sense.

The compiler thus rejects an attempt to redefine a name. If a namecode collision did occur, the same rejection would prevent its entry into the nametable. Thus the IPS name coding technique, despite its statistical reasoning, leads always to unique name codes.

The routine $NAME now needs a mechanism to map the original name space into 4 bytes. In order to solve the above-mentioned concatenation problem, the first byte is used to store the length of the name. With a maximum namelength of 63 characters, only 6 bits are needed. The remaining 2 bits are used to store the entrytype, as explained with the compiler routine.

The remaining 3 bytes contain the namecode proper. Subdivision of the name space into subspaces with fixed namelengths shifts the collision probabilities slightly.

A detailed analysis revealed that this approach reduces the collision probability with short names at the expense of very long names.

This is a desirable feature since short names are more common. In fact, with names of 3 letters or shorter, collisions turn out to be impossible.

The pseudorandom mapping process used is a division by a 24 bit polynomial. The remainder of this division is used as the code. $NAME does this division using all characters of the name. If it is inadvertently longer than 63, the loop doing the division nevertheless uses only the first 63 characters.

Before starting the division, the remainder is preset to some more or less arbitrary number, #57CE8D.
The reason for this is explained later.

The actual division is performed by the code routine $POLYNAME . It is organized to perform the operation using one character on the stack and modifying the remainder, also on the stack.

It operates mostly byte-wise and thus is very fast; so this complicated operation does not slow down the translation process.

Finally, $NAME stores the thus computed namecode in the field $ND (4 bytes) for further processing by other definitions.

```
$POLYNAME algorithm
   1. Get word and 1 byte [A B] C,
      and 1 character byte D from the stack
   2. Form 24 bit word X = [ A B C ]
   3. Calculate the product P:
        P = D EOR X EOR X>>1 EOR X>>2 EOR X>>7
           where X>>n means X logically shifted
           right n places
   4. Take the LS byte of P and form 32 bit word
      X = [ P A B C ]
   5. Shift X right 7 places
   6. Return three LS bytes of X = [ .. A' B' C'
      ] to stack, in entry format.
```

Some IPS versions use as the scanning routine ($CSCAN) a code routine because the compiler spends a significant fraction of the time in this scanner.

This is somewhat involved, because a scan may take more than 20 ms. Thus, provisions are needed to handle the pseudo-interrupt in between.

In the interest of portability the high-level $CSCAN is more straightforward and thus was selected for the faster processor.

5.2.2 Nametable and searching

In order to allow the compiler to communicate with the user, some relationship between the names (namecodes) and the objects thus identified must be established.

The names could be stored in front of the entry proper in memory. In this case a linking pointer has to be included with each name to facilitate searching.
This technique is called a "linked list".

Or, with a different approach, the names are stored in a separate table. In this case each name entry needs a pointer to point to the respective entry. Such a nametable could be organized in various forms.

One of the most effective approaches is to try to calculate from the namecode a position in the table. If this position does not contain the name, it may either be empty or contain a different name. If it is empty, the name is not in the table; if a different name is found, a recalculation of the position yields a new spot.

This may be continued until the name is found or an empty spot is hit.
This technique of list organization has become known as a "hash table".

Both techniques have advantages and disadvantages in the context of IPS.

The linked list is more open ended and makes slightly better use of the memory space. But the searching process typically has to look up more than one half of the entries until the name is found. This makes it essential that the search routine is a code routine to get an acceptable compiler speed.

The hash table keeps names separate from the program and thus is more suitable if objects are to be "transcompiled" for processors without human interaction peripherals. Its main disadvantage is the fixed size of the hash table.
Even if it is nearly empty, it has to be loaded in its entirety taking some time. But the search process is extremely fast; on the average only 1.5 accesses are necessary with a table half full. Thus the search algorithm may be coded in high level IPS without disadvantage. This helps portability of the system, of course.

At the time of the writing of this book, there exists only a linked list version of IPS for the COSMAC, but there are hash table versions in existence for all four processors. Thus, the hash table version was selected for this book.

The size of the hash table was chosen for 512 entries. Experience has shown that with a 16 Kbyte system table and the rest of the memory saturate at about the same time. Further, there are hardly any applications conceivable, where this size would be insufficient.

Each entry consists of 6 bytes, four for the name and 2 for a pointer to the actual routine. Thus 3 Kbytes of memory are needed for the table. The last entry
is not used; it contains the synch-vector for the tape cassette recording program.
Because the tape output buffer is supposed to follow the hash table, this simplifies the recording program.

To calculate the position of the name in the table, a random mapping technique is required. Since the namecode already represents such a random number, it suffices to take 9 bits of this code as a key to the table. With short names this would lead to some clustering.

For this reason, the namecode is preset to some non-zero value to ensure a good random distribution of the 9 bits even with short names.

If the position is in use by a different name, a new position must be calculated. This is done by the so-called quadratic rehash technique. Normally only half of the table could be reached by this technique.

If an odd increment (not the customary 2) is chosen, the whole table may be accessed. For a detailed discussion of hashing techniques please refer to the literature [2,3,4].

The definition $SUCH returns the pointer of the routine and on top of it the position of the first byte of the namecode. The compiler may inspect this to find the entrytype of the entry. If the name is not available in the table, $SUCH returns a 0 and on top of it the position where an entry with such a name may be placed into the nametable. If the table is full and the name has not been found, it returns two zeros.

5.3 Compiler auxiliary routines

The other routines called by COMPILER are straightforward and thus are described only shortly.

$ZAHL uses a word bracketed by $P1 and $PI (provided by $NAME) and tries to convert it to a number. It first checks if the first character is - # or a B and sets the conversion base accordingly. The sign is pushed on the return stack.

Then the conversion proceeds character by character. $ZAHL returns the number and on top of it a 1 if the conversion found only characters consistent with the number base; else it returns an undefined number and on top of it a 0.

$DEP deposits a two byte number from the stack at HERE and then increments $H by two.

$CEN is the clean-up operation at the end of input processing. It expects on the stack the position of where the new input is to go and from where it is to be processed.
It sets the input pointer, places a cursor there and sets $PI . It puts $TVE into $PE and then clears the READYFLAG .

IE is the termination if an error has been found in processing the input. It disables the cassette reading by clearing TEXTREAD and then places a ? in front
of the erroneous word. IE then calls $CEN with this position of error to give the user the opportunity to correct the error. IE takes one number off the stack and discards it.

SYSWRITE expects the address of a 16 byte long field on the stack containing an error message. It writes this message to the TV screen and then terminates further processing by calling IE .

L>>> checks the length of the field to be transported and if necessary breaks this transport down into consecutive calls of >>> . With the length equal or smaller than zero no transport takes place.

5.4 Entry generating routines

Entries are created, among others, by : , VAR , CON or FIELD . These all call first ENTRYSETUP to place an entry into the nametable. ENTRYSETUP returns a 1 if this was successful. HERE points to the parameter field of the entry and below the 1 on the stack there is the pointer to the typecode field of the entry.

If ENTRYSETUP could not create an entry because no name was in the input, it writes NAME MISSING and returns only a 0 . It then stops processing by calling IE .

The creation of a variable, a constant or a field is straightforward. The parameters returned by ENTRYSETUP are used to complete the entry. Into the typecode field are placed the starting addresses (code !) of VAREX or CONSEX , the executive code for variables and constants.

The creation of a : entry involves some more steps. After calling ENTRYSETUP the typecode field is filled by DEFEX , the executive for definitions.

The COMPILEFLAG is set and the address of $F2 is pushed on the stack. This last action serves two purposes.

First, the word ; checks and removes this number from the stack. If there has been a structuring error, there is either a residual address additionally on the stack, or this number has been removed by one of the jump words by taking it as an address. Thus structuring errors are identified.

The second reason for this number is also connected with structuring errors. If a word like YES? has been omitted, the THEN will treat any number on the stack as the location to insert the jump address. If a number is on the stack that is an address of vital code, a structuring error could bomb the system.

By placing $F2 on the stack, this address is placed into $F2, a variable just created for this purpose. This is also the reason why the first underflow position of the stack is preset by $F1 ; it serves the same purpose and still allows the distinction between $F2 to be able to identify a structuring error.

Entry summary:

An IPS entry consists of four fields:

```
Namecode field            4 bytes
Pointer or Link field     2 bytes
Typecode field            2 bytes
Parameter field           n bytes
```

Namecode example: **:INT WEG/AB** encodes as `#C6 #D5 #10 #17`.

The first byte is #C0 (entrytype :INT) plus 6 (number of bytes in the name).

The other three bytes are produced by $NAME running "WEG/AB" through $POLYNAME . (Sect. 5.2.1)

Pointer or Link: Namecodes are collected into either a nametable (4+2 = 6 bytes per entry) or a linked list. In a nametable system the pointer points to the entry's associated typecode field.
In a linked list system, Link points to the namecode field of the previous entry; the first entry has the value of Link set to zero. (Sect. 5.2.2)

Typecode: The address of executable code that characterises an entry.
Standard typecodes are CON VAR : and FIELD (as above).

Parameter field: Additional data required by the executable code; for example the actual 2-byte value of a variable (VAR) entry. A definition (:) entry is followed by many bytes, representing the words that make up the definition.

Now let us have a closer look at ENTRYSETUP . It first calls $NAME to get the name of the entry. It then procures the position in the nametable where this entry may be placed by calling $SUCH.
If an entry with this name already exists, processing is stopped. (This feature may be disabled by clearing $RS .)

Processing is also terminated if the table is full. Otherwise, ENTRYSETUP proceeds by transferring the name from $ND into the nametable.
Because ENTRYSETUP is also used for colon entries, it stores the position of the entry in the nametable into $KK .
If the entry needs to be removed later due to a structuring error, this position is available to restore the table to its original state.

Storage of the table position also enables the routines calling ENTRYSETUP to modify the entrytype bits in the name if a different than normal entry is generated.

In order to prevent damage to the program if ENTRYSETUP was unsuccessful, $KK is set to $F1 in this case. An entrytype modification attempt will then take place in $F1 and thus causes no damage.

The word ; deposits the address of RETEX (the definition return executive) and then reverses the action of the : .
If it did not find $F2 on the stack, it eliminates the entry using the table location stored in $KK .
It then terminates processing with an error message.

Definitions of types other than normal are produced by the words :HPRI , :PRIOR and :INT . These simply call : and afterwards modify the type in the name by calling PRIMODIFY .

This routine uses the position of the name in the table as stored in $KK to set the appropriate bits.

5.5 Structuring words

The control of flow in the program is governed by the branch generating words like YES? NO: THEN and LBEGIN LEND? . In addition to depositing the appropriate branch instruction these words also have to handle the jump addresses.

The procedure is identical to the way the assembler handles this. The jump addresses or the positions where the jump addresses are to be inserted later, are kept on the stack.

All entries of this sort are of type :PRIOR because they are executed during the compilation.
In the interpretive mode they are illegal.

The description of the machine dependent part contains a list of all instructions (code routines) related to branch operations.

5.6 Address fetching

Writing programs in IPS involves the fetching of various addresses. With variable entries this is automatic, mentioning the variable name pushes the address of the parameter field on the stack.

With constants the ? may be used to get the address of the parameter field.

On other occasions, particularly with systems programming, the typecode field is required. The word ' followed by the name pushes this address on the stack.

The word ' is of type :HPRI . This means that it is executed in the interpretive mode and in the compile mode. In compile mode (COMPILEFLAG set) this word first deposits the code of 2BLITERAL and then the address. Thus the address is pushed on the stack only at execution time of the definition.

Most of the address fetching routines use the word $GETADR . This definition first calls $NAME to encode the name of the word for which the address is to be fetched. Then $SUCH is used to locate the address. $GETADR returns a 1 if successful and below it the address. Otherwise it returns only a 0 and also terminates input processing.

5.7 Text handling routines

A set of routines is provided to facilitate text manipulation. Central towards this end is the definition " allowing direct entry of text, either as string to be entered into fields or as literal operation in definitions.

The :HPRI word " first increments $PI over the first blank after the " and then scans the input string using $PI until it finds another " or it has scanned over the end of the TV screen. The length of the string minus one (to account for the blank required in front of the terminating ") is checked to be in the range of 1 and 256.

If not, processing is halted with a message. Provided that the length is ok, further action depends on the COMPILEFLAG .

In the interpretive mode, " returns simply the address of the first character of the string on the screen. If IPS is in compile mode, at first the address of TLITERAL is deposited. The following byte contains the length of the string.
Finally the string is deposited, HERE pointing to the first byte after the string.

At execution time TLITERAL , being a definition, finds the pseudo-program counter on the return stack. It thus can fetch the length of the string and the string itself and write it on the TV screen using SP as pointer.

Also the return stack containing the pseudo-PC is incremented beyond the string. Thus, the program will continue with the first instruction following the string.

The text editing mode finally is entered by setting the EDITFLAG to 1. The 20 ms keyboard handling routine then does not set READYFLAG after a return and enables additional editing keys. The edit mode thus is handled within the 20 ms pseudo-interrupt.

5.8 Stack display

The definition STACKDISPLAY is normally in the chain at location 0. If the compiler reached the end of the input buffer after having processed the last input word, the variable $P2 is set to 1.

The next pass through the compiler clears it again. STACKDISPLAY picks up this variable and displays the stack contents if $P2 is set.

At first, the first two lines of the screen are cleared to make room for the new display. Then the stack pointer is fetched by calling $PSHOLEN . (With the 6502 it is picked up directly from memory.)

By subtracting $SL (stack limit), the number of stack items is calculated. If it is more than 32 bytes (16 numbers), only the 16 top most are displayed.

The number of bytes is put on the return stack, the parameter stack contains the address of the topmost entry. A loop is entered now using the current number as index for the TV screen display position.

The stack content is then picked up. With the COSMAC and the 6800 the stack maintains the most significant byte at the lower address. It thus has to be transposed, before it can be displayed using CONV.

This routine uses SP. To prevent clobbering SP with the stack display, SP is pushed on the return stack and at the end of the display it is restored again.

The definition CONV is straightforward; it uses BASE , the variable used by the input number converter $ZAHL , as its number base. Thus the display format depends on the type of the last number entry. It works by continuously dividing the number starting with the highest possible digit value (10,000 or #1000) and then displaying the quotient.

The remainder then is divided by the next digit value and so on. With the decimal display mode leading zeros are suppressed by keeping a switching variable on the return stack. If a zero is converted, at least
one zero is displayed.

5.9 Chain control operators

Since IPS was primarily designed for real-time control applications, some provision is needed to let the computer handle several tasks "simultaneously".

The processor of course can serve only a single task at a time. Therefore, some strategy is needed that distributes the available processor time in a fair way between the tasks. chain principles

A typical solution in large scale processing centres would be to give to each task a certain fraction of the total time available and switch between the tasks often enough so that no task is excessively delayed.

Unfortunately, this scheme would be grossly inefficient in the problem area we are concerned with, because in real-time applications most of the tasks usually are dormant (waiting for some external or internal event). Thus, task oriented time slices would allocate most of the time to waiting tasks.

The problem may be resolved by the observation that if a task becomes active, it usually takes only very little time to service it until it becomes dormant again. This suggests serving tasks in a "round robin" fashion.

If the tasks are put in a ring, the processor polls each task to see if it needs servicing. If not, the processor is passed to the next task. A task needing service then may grasp the processor and do the required actions before passing on to the next task.

This is an effective solution as long as single tasks do not occupy the processor for an excessive amount of time. Since this is under control of the programmer, this is not a serious restriction.

Solutions of this kind have been proposed in [5] and [6]. The straightforward implementation of this technique assigns to each task individual stacks (parameter and return).

Most 8-bit processors typically would maintain some or all system pointers in on-board registers. This results in significant context swapping overhead as the program jumps from task to task. These routines must be implemented at machine level and thus are not portable. Worst, keeping two stacks per task would lead to an undesirable fragmentation of the memory space. Particularly with the small systems for which IPS was created, this is a serious drawback.

This situation led to the question of whether the tasks really need individual stacks. With the ring structure proposed (chain) context switching occurs only under control of the programmer. Interrupts do not present a problem because they do not communicate with the programs in the ring via the stacks and thus do not modify them as far as the chain is concerned.

Using only one parameter and one return stack shared by all tasks naturally leads to some restrictions:

if a task is abandoned for the time being, it should leave no items on the stack. Practical experience shows that this normally would be the case anyway, and where not, no problems result from satisfying this rule.

A single return stack is more serious. It implies that the suspension and resumption of programs in the chain can occur only at certain levels within the nesting of definitions. Only practical experience could tell whether this is overly restrictive. After two years of experience with the shared stack solution nosituation was encountered yet where this technique was found to be inadequate or too cumbersome and involved.

As IPS is put to more varied uses, a more general solution may become necessary, though.

The chain technique described is implemented in addition to the chain itself by four operators.

The words DCHN and ICHN are straightforward as to their implementation. The chain contains the address of the definition constituting the task.
The word $INSERT receives the chain location to be modified (0-7) and on top of it the address to be put there.

$CHAINACT uses this word to perform the chain modification. If IPS is in the interpretive mode, the chain location is checked to be in the range 0-7. Otherwise an error stop is performed.

In compile mode the action is to be performed later upon execution of the definition. Thus the address must be compiled as an immediate instruction before compiling the $INSERT.

DCHN fetches the address of the no-op (RUMPELSTILZCHEN) and inserts it into the chain.

ICHN on the other hand gets the address of the following word and puts this into the chain.

The words WAITDEF and YES:CONT require a storage area for storing the original chain content and the address as to where the program is to resume.
The field IMMAGE thus has four bytes for each chain position.
The word WAITDEF stores the chain address and the resumption address in this field.

When called it first discards its own return address, then calculates the chain location it involves by making use of the fact that the third item on the return stack is this address plus two. It then stores the chain content of this location and the resumption address (item 2 on the returnstack) at the appropriate location in the field IMMAGE.

YES:CONT reverses this process if it finds a 1 on the stack. It again discards its own return address, again calculates the chain location, picks up from IMMAGE plus chain position times two (the first index) the address to continue, pushes it on the return stack and then picks up from IMMAGE 16 bytes higher the original chain operator and stores it in the chain.

IMMAGE contains, in the 16 lower bytes, the resumption address and in the 16 higher bytes the original chain operators.

5.10 Miscellaneous high level routines

The word ? fetches the address of the parameter field of a constant. Because $GETADR fetches the address of the typecode field, this address must be increased by two. ? is defined as :INT to prevent its use in definitions in view of ROM versions of IPS.

$LOAD initializes a load operation by setting the load limit pointer and the running load pointer. Then it sets LOADFLAG thus starting the load sequence and decoupling the tape input from the TV/keyboard input.

READ starts a load sequence into the tape buffer $BU .

RECORD starts the recording action by putting #70 into the variable C/Z . The 20 ms routine picks it up there and starts the recording. After completion, it is reset to zero by the 20 ms routine.

DEL/FROM eliminates an entry and all subsequent ones from the nametable. If the address of the entry is smaller than $LL (the beginning of the user area), the operation is illegal because it is assumed that the system is not to be removed.

Provided that the removal is legal, $H is reset and all table entries having equal or larger addresses than the one to be removed are cleared.

6 The machine dependent routines

The low level routines of IPS break down into the IPS words defined by machine code (the primitives) and into the executive responsible for bringing these routines to execution and doing various housekeeping tasks.

The IPS codewords are made up of two groups.

The words constituting the high level are already known from the description in chapter I.

In addition, there are other words that are not directly handled by the user, but are compiled in the right context by the compiler. (Like the branch compiled when the compiler encounters a YES?).

6.1 More details about the executive

Most of the executive is described in section 4. Please refer there for its operation. With the COSMAC the 20 ms routine is completely implemented as a pseudo-interrupt and the tape output is handled by DMA hardware.

With the other processors the 20 ms actions occur in two steps. The sequence starts with an interrupt every 20 ms. Within this interrupt two actions are taken: the tape output is served if necessary and the emulator is modified in such a way as to jump to the 20 ms pseudo-interrupt routine after completion of the current code routine.

The 20 ms routine performs these tasks: IPS executive

1. Serve an A/D converter, if available.

2. Handle tape recordings, if not already done by the true 20 ms interrupt, including motor control.

3. Check, in load mode, if the load pointer exceeds the INTLIMIT.
   If so, reset LOADFLAG.

4. If not (READYFLAG or EDITFLAG), check if the input pointer exceeds $PE, if so set READYFLAG.

5. Enable the cassette read motor if not (TEXTREAD and READYFLAG) or if LOADFLAG ; else disable motor.

6. Transport output bytes from field to ports; input bytes from ports to fields where applicable.

7. Update clock and stopwatches

8. Input keyboard character, if available. If READYFLAG not, the character processor is called. This processor first normalizes the input pointer to ensure it is on the screen and then removes the cursor.
   Unless some special control character is found, the character is put on the screen and the input pointer is incremented.
   Then the cursor is set again.
   If an editing character is found, the appropriate action is erformed.
   If EDITFLAG is set, the character processor checks for additional control characters and

performs the action (including the carriage return). If EDITFLAG is not set, the carriage return results in $PE set to the input pointer minus 1; the cursor is not set in this case.

The input limit handler (point 4) finds this situation and sets READYFLAG . (This limit also exceeds $PE if a tape input has occurred and EDITFLAG is not set. Thus the equivalence of keyboard and tape input).

The special editing functions taking more time than 20 ms may call the items 1 - 7 while performing the action. The character processor again normalizes the input pointer before leaving the 20 ms routine.

6.2 Directly available code routines

The following words implement functions introduced in chapter I:

```
@     @B    !     !B    DUP   SOT    DEL     SWAP   RTU
I     R>S   S>R   =0    <0    >0     INV     AND    OR
XOR   BIT   >>>   + -   *     /MOD   F-COMP  RUMPELSTILZCHEN
```

The following words may also be used by the user directly. Their function is somewhat specialized and they are thus not necessarily very useful outside the compiler.

Therefore, they are defined as part of the implementation rather than part of the language IPS. Not all of them are available in all installations.

RETEX exits from a definition.
It is compiled by the ; and using this word directly allows multiple exits from a definition.

$TUE expects a typecode field address on the stack.
$TUE executes this word and thus allows indirect execution.

$IPSETZEN takes a number off the stack and puts it into the input pointer.

$LPSETZEN takes a number off the stack and puts it into the load pointer.

$PSHOLEN gets the parameter stack pointer and pushes it on the parameter stack.

$PSSETZEN takes a number off the parameter stack and sets the stack pointer to this number.

$POLYNAME divides a character by the name encoding Polynomial (24-bits) and leaves the remainder on the stack.
It expects the previous remainder on the stack and on top of it the character (3 numbers).
It leaves 2 numbers, of which one byte is unused.
See Sect. 5.2.1 for the details.

6.3 Code routines managed by the compiler

The following code routines can only be compiled within the proper routines; trying to execute them from the keyboard may result in a system crash.

2BLITERAL within programs pushes the number (2 bytes) following it on the stack. The pseudo-PC is incremented beyond these 2 bytes.

1BLITERAL like 2BLITERAL, but pushes only 1 byte following in the program (0 - 255) on the stack and then increments the pseudo-PC beyond the byte. (These 2 words correspond to the "immediate" instructions of assembler programming)

BRONZ branches on finding a 0 (even number) to the address following it in the program.
If it found an odd number,
it increments the pseudo-PC beyond the address.

LOOPEX the action taken with the NOW of loops.
The address to the loop entry follows LOOPEX in the program.

+LOOPEX like LOOPEX, but for the +NOW taking the increment off the stack.

$JEEX The action going with the EACH . The address of the corresponding LOOPEX address follows $JEEX in the program to enable it to bypass the loop in case it s not to be executed at all.

JUMP like BRONZ, but the jump is always taken and nothing is taken off the stack. (Unconditional jump).

The following three routines define the types of IPS entries (definitions, constants and variables including fields); their starting addresses are put into the typecode fields of entries rather than into the normal program instruction stream.

Thus, they do not require the header of the other code routines. For the compiler they are represented as constants containing the starting address of the code.

The compiler places this address into the typecode field of the entry when creating it. Upon execution of the entry, this address is found in the second step of the emulation process and thus it must point to executable code.

DEFEX definition execution; pushes the pseudo-PC on the return stack and put HP +2 into the pseudo-PC.

VAREX variable execution; pushes HP +2
(the parameter field address) on the stack.

CONSEX pushes the content of the memory pointed to by HP +2 on the stack.

6.4 Initialization

In addition to the executive described, all IPS versions require an initialization routine setting up the various pointers and registers before the emulator can be entered.

This routine is only run through once; the memory space can be used for other purposes later on.

7 Typical IPS Compiler Routines

This generic IPS compiler and associated routines are based on the meta-language coding of IPS-C, linked list version. It is designed to be processed by a cross-compiler.

When compiled, lower-case words will become uppercase, destined for the target processor IPS image.
Words :n and ;n become : and ; .

This listing uses German words as discussed in Sect. 5. Correspondence between words is given below.

Only differences are shown. Words ~ 'n +n 02 !t and $OC are cross-compiler directives.

+NUN	+NOW	NEIN:	NO:
ANFANG	LBEGIN	NUN	NOW
BASIS	BASE	ODER	OR
DANN	THEN	PWEG	PDEL
DANN/NOCHMAL	THEN/REPEAT	RDU	RTU
ENDE?	LEND?	UND	AND
EXO	XOR	VERT	SWAP
FELD	FIELD	WEG	DEL
HIER	HERE	ZWO	SOT
JA?	YES?	BEA	LIS
JE	EACH	BEM	LIM
KON	CON	EINGABEZAHL	N-INPUT
NICHT	INV	Z-LESEN	N-READ
		WEG/AB	DEL/FROM

SPEICHER VOLL !	MEMORY FULL !
NAME FEHLT !	NAME MISSING !
STAPEL LEER !	STACK EMPTY !
STRUKTURFEHLER !	STRUCTURE ERROR!
TEXTFEHLER !	TEXT-ERROR
UNZUL. NAME !	DUPLICATE NAME !

```
                ( Typical Compiler Routines )
                    ( Based on IPS-C )

#0004 feld $ND                  #0001 var $RS
#0000 var $F1                   #0000 var $F2
#0000 var $KK                   #0000 var BASIS
#0000 var BEM                   #0001 var BEA
#0000 var EINGABEZAHL           #0000 var Z-LESEN
#0000 var   COMPILEFLAG         #0000 var $V1
#0000 var   LINK
16 feld     STACKMESSAGE      16 feld   MEMMESSAGE
16 feld     NAMEMESSAGE       16 feld   STRUCMESSAGE
16 feld     TEXTMESSAGE       16 feld   RSMESSAGE

~ SPEICHER VOLL !  ~ 'n   MEMMESSAGE   02 +n $OC   !t
~ NAME FEHLT !     ~ 'n   NAMEMESSAGE  02 +n $OC   !t
~ STAPEL LEER !    ~ 'n   STACKMESSAGE 02 +n $OC   !t
~ STRUKTURFEHLER ! ~ 'n   STRUCMESSAGE 02 +n $OC   !t
~ TEXTFEHLER !     ~ 'n   TEXTMESSAGE  02 +n $OC   !t
~ UNZUL. NAME !    ~ 'n   RSMESSAGE    02 +n $OC   !t

 ( DEFINITIONEN )
:n HIER    $H @ ;n
:n H2INC HIER 2 + $H ! ;n
:n $DEP    HIER ! H2INC ;n
:n $CEN    $BU $PI ! 0 READYFLAG !B ;n
:n IE    $P1 @ $PI @ 1 - je I @B #80 ODER I !B
                        nun 0 $P2 ! $CEN WEG ;n
:n SYSWRITE SYSLINE 16 >>> 0 IE ;n
:n L>>> anfang DUP 256 > ja? 256 - S>R PDUP 256 >>>
                            256 + VERT 256 + VERT R>S
        dann/nochmal DUP >0 ja?   >>>
                           nein: PWEG WEG
                           dann ;n
code $SCODE ( Machine specific code for $SUCH. )
            ( see Chapter III sect. 5.1        ) NEXT
code $CSCAN ( Machine specific code for $NAME. )
            ( see Chapter III sect. 5.2.1      ) NEXT
```

```
:n $SUCH LINK @ $P3 ! $SCODE ;n
:n $NAME   0    READYFLAG @B
       ja? 1 $CSCAN >0
          ja? $PI @ $P1 !
               2 $CSCAN PWEG #CE57 #8D
               $P1 @ $PI @ ZWO - DUP 63 > ja? WEG 63
                                           dann
               DUP $ND !B 1 - ZWO +
               je I @B $POLYNAME
               nun $ND 3 + !B $ND 1 + ! 1
          dann
       dann ;n
:n $ZAHL 1 ( OK ) 0 ( ANF. ) $PI @ 1 - $P1 @
  #2D ZWO @B = ja?     1 + -1 S>R ( NEG ) 10 ( BASIS )
               nein:         1 S>R ( POS )
                     #23 ZWO @B =
                     ja?   1 +               16
                     nein: #42 ZWO @B =
                           ja?    1 +          2
                           nein:              10
               dann dann dann     BASIS !
 VERT je BASIS @ * I @B DUP #3A < ja? #30 -
                                  dann
                         DUP #40 > ja? #37 -
                                  dann
   DUP BASIS @ >= ZWO <0 ODER ja? ( FEHLER ) WEG 0 RDU
                                     dann +
      nun R>S * VERT ;n
:n COMPILER $NAME
ja? $SUCH
    1   ( For continuing ) BEM @B
        ja? ZWO 'n RUMPELSTILZCHEN
               = ja?   ( RUMP. ) 0 BEM !
                 nein: ( NICHT RUMP. ) Z-LESEN @
                      ja?   PWEG 0 1
                      nein: ZWO BEA @ <
```

```
                    ja? IE WEG 0
                    dann
                dann
             dann
     dann    ( End limited input mode checks )
```

```
    ( COMPILER cont. )
    ja? ( Continue flag ? ) DUP =0
        ja? ( Number processor )
          WEG $ZAHL
           ja? COMPILEFLAG @B
            ja? DUP #FF00 UND
              =0 ja? 'n 1BLITERAL $DEP
                         HIER !B $H INCR
                  nein: 'n 2BLITERAL $DEP $DEP
                  dann
            nein: BEM @B ja? EINGABEZAHL ! 0 Z-LESEN !
                             dann
             dann
           nein: IE
           dann
      nein: ( Found processor ) DUP 6 - @B #C0 UND
             COMPILEFLAG @B ODER
             DUP 1 =
             ja?    WEG HIER $ML >=
                           ja? WEG MEMMESSAGE SYSWRITE
                                nein: $DEP
                                dann
             nein: DUP #80 = VERT #C1 = ODER
                    ja?      IE
                    nein: R>S $V1 ! $TUE $V1 @ S>R
                    dann
             dann
      dann
      $PSHOLEN $SL > ja? $SL $PSSETZEN
                         STACKMESSAGE SYSWRITE WEG $F1
                      dann
   dann
dann READYFLAG @B $P2 @B UND
     ja? #20 $BU !B $BU DUP 1 + 511 L>>> $CEN
     dann
;n
```

```
      ( Compiler Auxiliary Routines.     )
             ( See Sect.5.3 & 5.4 )

:n ENTRYSETUP $F1 $KK ! $NAME DUP
     ja? $SUCH =0 NICHT $RS @ UND
          ja?   RSMESSAGE SYSWRITE WEG 0
          nein: HIER DUP $KK ! LINK @ H2INC H2INC $DEP
                $ND ZWO 4 >>> LINK ! HIER VERT H2INC
          dann
     nein: NAMEMESSAGE SYSWRITE
     dann ;n

:n $GETADR $NAME ja?        $SUCH DUP =0
                                  ja?    IE     0
                                  nein:         1
                                  dann
                       nein: NAMEMESSAGE SYSWRITE 0
                       dann ;n

:hpri    '        $GETADR ja? COMPILEFLAG @
                          ja? 'n 2BLITERAL $DEP $DEP
                             dann
                         dann ;n

:prior       ;      'n RETEX $DEP 0 COMPILEFLAG !B
             $F2 <>
             ja? STRUCMESSAGE SYSLINE #20 + 16 >>>
                 LINK @ DUP $H ! 4 + @ LINK !    0 IE
             dann ;n

:int     :    ENTRYSETUP ja?
                     DEFEX VERT ! 1 COMPILEFLAG !B $F2
                         dann ;n

:n PRIMODIFY         $KK @ @B ODER $KK @ !B ;n

:int :PRIOR          :    #80 PRIMODIFY ;n

:int :HPRI           :    #40 PRIMODIFY ;n

:int :INT            :    #C0 PRIMODIFY ;n
```

```
:prior   JA? 'n BRONZ $DEP HIER H2INC ;n
:prior   DANN HIER VERT ! ;n
:prior   NEIN: 'n JUMP $DEP HIER H2INC VERT DANN ;n
:prior   JE    'n $JEEX $DEP HIER H2INC ;n
:prior   NUN   'n LOOPEX $DEP DUP DANN 2 + $DEP ;n
:prior   +NUN 'n +LOOPEX $DEP DUP DANN 2 + $DEP ;n
:prior   ANFANG HIER ;n
:prior   ENDE? 'n BRONZ $DEP $DEP ;n
:prior   DANN/NOCHMAL VERT 'n JUMP $DEP $DEP DANN ;n

:int KON          ENTRYSETUP ja? CONSEX   VERT ! $DEP
                                dann ;n

:int VAR          ENTRYSETUP ja? VAREX    VERT ! $DEP
                                dann ;n

:int FELD         ENTRYSETUP ja? VAREX    VERT ! HIER + $H !
                                dann ;n    FIELD   CON

:n !CHAR      SP @ !B SP INCR ;n
:n TLITERAL I 1 + R>S @B PDUP + S>R SP @ PDUP + SP !
           VERT >>> ;n
:hpri "
   $PI INCR $PI @
   anfang $PI @ DUP @B #22 = VERT $RBE > ODER $PI INCR
    ende?
    $PI @ 2 - ZWO - DUP 1 < ZWO 256 > ODER
    ja?   TEXTMESSAGE SYSWRITE VERT WEG
    nein: COMPILEFLAG @ ja?
                        S>R I 'n TLITERAL $DEP
                        HIER !B $H INCR
                        HIER I >>> HIER R>S + $H !
                       dann
    dann ;n
:int !T VERT >>> ;n
         ( End of Compiler and Auxiliary routines)
```

Appendix A

IPS - Short Reference Summary

' Words ' are arbitrary character strings.
They are separated by at least one blank, by comments or by 'end of block'.

`( ... )`	`Comment`
`number CON name`	`Constant definition`
`number VAR name`	`Variable definition`
`n FIELD name`	`Definition of a field of n bytes length`
`: name ... words ... ;`	`Creation of a program definition`

Text Editor:

(entry by TEXT , exit by control D (EOT))

(Also available during IPS-entries)

`control \ ( FS )`	`Forwardspace`
`control H ( BS )`	`Backspace`
`control J ( LF )`	`Linefeed`

(Only available in edit-mode)

`control P ( DLE )`	`Deletes all after cursor, cursor to top`
`control M ( Return )`	`Normal function, not IPS entry indication`
`control ^ ( RS )`	`Recording start unction`

(Not with all IPS versions)

control O	Deletes character
control N	Inserts blank for character insertion
control R	Deletes line
control Q	Inserts empty line
control X	Set blockcounter (1-5) to 1
(DEL)	Move cursor one line up

1. Stack manipulations

	Before operation	After operation
DEL	a b	a
PDEL	a b c	a
DUP	a	a a
PDUP	a b	a b a b
SWAP	a b	b a
SOT	a b	a b a
RTU	a b c	c a b
RTD	a b c	b c a

```
S>R    Stack to return stack           )
R>S    Return stack to stack           ) one number
I      Duplicate return stack to stack )
```

2. Arithmetic and logic operations

+		AND	and all bits
-		OR	or all bits
*		XOR	exclusive-or all bits
/		INV	complements all bits
MOD	remainder	BIT	select one bit 0 - 15
/MOD	quotient, and remainder on top		

3. Comparison operators

=0 >0 <0 = < > =/ >= <=

F-COMP (expects two addresses and length in bytes. Compares two arrays and delivers 1 if equal, 0 with smaller and 2 with larger)

4. Storage operations

`@`	Get word
`@B`	Get byte; MS-byte =0
`!`	Store word
`!B`	Store byte; LS-byte only, MS-byte discarded
`!T`	stores text from screen. Format: " text " fieldname !T
`!FC`	stores field components. Format: **`n1 n2 n3 ... nZ          fieldname  Z  !FC`**
`!CHAR`	stores a number as character as pointed at by SP and then increments SP

?name gets address of constant to store a new value into it

INCR expects variable address; increments its content

5. Control of flow

a. **YES? NO: THEN**
b. **EACH NOW or +NOW**
c. **LBEGIN LEND?**
d. **LBEGIN YES? THEN/REPEAT**

6. Field operations

>>>	from-addr to-addr num-of-bytes (1-256) >>>
L>>>	same as >>> , but length 0 - 32767 bytes
!FC	see Storage operations, 4.
!T	see Text operations, 7.
TRANSPORT	expects address, moves 512 bytes starting at this location into tape output buffer $BU

7. Text operations

" arbitrary text-string not containing quote "

within definitions: text will be written at execution time starting at SP, SP then points at first unused position.

outside definitions: leaves on stack address of first character (on TV) and length of string on top. Length of string is number of characters between quotes minus two.
This may be 1-256 characters.

!T	stores text strings in fields. Field must have sufficient length. Format: **" ... " fieldname !T**
BLANKS	expects a number n; it writes n blanks starting at SP . SP is not modified.
WRITE	expects fieldname and on top number of characters to be written. It writes starting at SP , SP is updated to next location after string.
CONV	expects a number. It converts it into a string and writes it at SP . SP then is updated as in WRITE . The conversion format depends on content of variable BASE .

8. Systems operations

DEL/FROM name	Deletes all entries starting at name.
READ	Reads a cassette-block into field $BU.
RECORD	Writes a cassette-block from $BU to tape.
OK	Clears error messages, enables cassette read.
n DCHN	(n=0-7) dechains program at position n.
n ICHN name	(n=0-7) puts definition 'name' into chain at location n .
' name WAITDEF	replaces current program in chain temporarily by definition 'name'. Only permissible at two definition levels below chain.
n YES:CONT	if n is odd, it reverses the action of WAITDEF and resumes the program with the word following WAITDEF. Only permissible at one Definition level below chain.

Note: WAITDEF and YES:CONT are to be used with empty stacks only.
Inside the EACH ... NOW there are items on the return stack.
Use LBEGIN ... LEND? with iteration parameters in variables instead!

STACKDISPLAY	Normally in chain at location 0.
HERE	Delivers first free position of dictionary.
RUMPELSTILZCHEN	No-op; in limited input mode it returns the computer to the normal mode.

9. Variables, constants and fields

IPS-system:

Name	Kind and length		Function
`CHAIN`	F	16	Main program. TV0 TV4 TV8
`$H`	V		Containing first free position of dictionary.
`TV0`	C		1st TV screen line position.
`TV4`	C		4th TV screen line position.
`TV8`	C		8th TV screen line position.
`$TVE`	C		Last TV screen position.
`C/Z`	V	1	Cassette recording control.
`TEXTREAD`	V	1	Tape read-enable flag.
`READYFLAG`	V	1	1 while there is still input in $BU to be processed.
`LOADFLAG`	V	1	1 while READ or $LOAD active.

Name	Kind and length	Function
BASE	V	Conversion base for CONV .
N-INPUT	V	Destination of number input in limited input mode.
N-READ	V	Number requested in LIM.
LIM	V	limited input mode flag.
LIS	V	Address of first word to be recognized in LIM.
SP	V	Screen pointer for text output.

Process control related:

Name	Kind and length	Function
OUTPUT	F 4	Hardware output ports) not with
INPUT	F 3	Hardware input ports) all
A/D-CONV	F 8	A/D-converter input) systems
CLOCK	F 6	0.01s, sec, min, hr, day-lsb, day-msb
SW0)		Stopwatches running backwards.
SW1)	F 4	0.01s, sec, min (2 bytes)
SW2)		After timeout 0.01s=1, rest=0.
SW3)		With 0.01s odd, watch is halted. IPS fields

10. Storage and reloading binary object programs

1. After program has been translated, translate DUMPER (3 blocks).
   It requires 74 bytes of storage, adds no names to dictionary and leaves one number on the stack.

2. Load cassette and enter $TUE . This starts the dump of total system (IPS plus translated program).

3. DUMPER is good only for one dump; if more are required, again translate dumper.

4. Program is loaded like IPS itself, it in addition contains the translated program in the state (chain, e.g.) at the time of the dump.

The source of DUMPER is lost. Though specifically for the Atari 800XL IPS and its 6502 processor, the following is similar. [Ed]

```
: DUMP HIER #FF ODER 1 + DUP #DD ! #1000 #DB !

       HIER #1234 !

       #1000 - 256 / #0A00 !B

       AUFNAHME

       ANFANG C/Z @B =0 ENDE? ( Wait loop )

       $BU #DB ! #0A00 #DD ! ; ( Restore pointers )
```

Appendix B

The AMSAT cassette standard

1. **Tape speed:** 4.76 cm/s (normal consumer recorder speed)
   Data rate : 400 bit/s

2. **Recording format** of the bits:

D	State of data (0 or 1)
Cl	400 Hz clock (squarewave)
4Clv90	1600 Hz squarewave
8Clv90	3200 Hz squarewave 90 degrees shifted with reference to Cl transition

$$Cs = ((D \oplus Cl) \,\&\, 4Clv90) + (\overline{(D \oplus Cl)} \,\&\, 8Clv90)$$

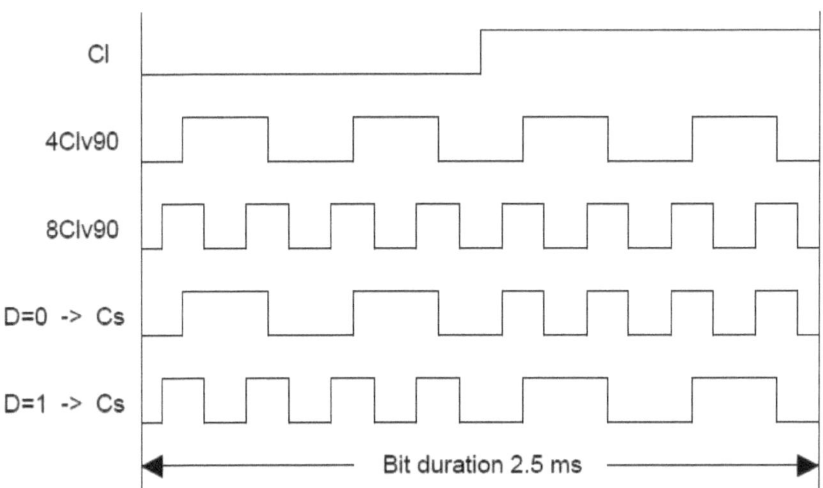

3. Structure of the recordings:

a. Byte format: bit serial, most significant bit first, least significant bit last.
Synchronous recording,
i.e. no start or stop bits.

b. Block format: (hexadecimal representation)

Blockbegin	Motor start
50,50,50,...,50,50,	64 bytes leader
39,15,ED,30,	Synchvector
data-byte0,db1,db2,...	Data field 512 bytes
...,db510,db511,	End of datafield
50,50,50,...,50,50,	48 bytes trailer
50,50,...	Motor stops here

Note 1: while motor comes to stop, the recording of 50's continues; there are no true block-gaps

Note 2: for recordings requiring exceptional data integrity, the first two bytes of the trailer may be a CRC .

4. Tapes:

Standard audio tapes (Fe2O3), C60 , C90 or C120. Two-track recording (mono).

5. Audio levels:

The levels comply with the DIN standards and connectors. In particular the recording interface produces approximately 15 mV p-p at pin 1 of the DIN connector for the recorder.

The output circuit contains some harmonic suppression to prevent beats with the bias-current of the recorder:

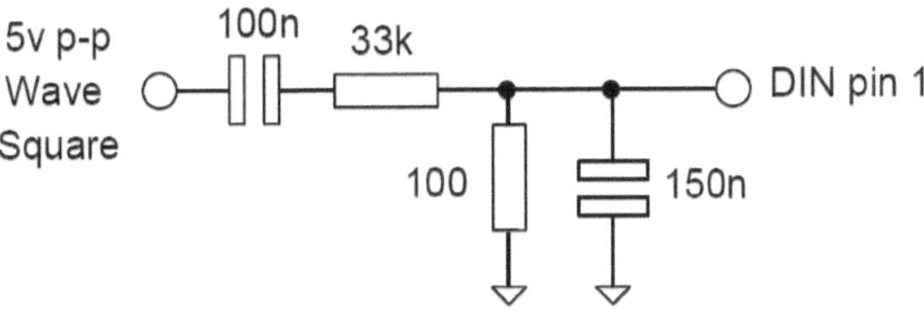

The reading interface expects about 2 V p-p at pin 3 of the DIN connector from the recorder. The internal resistance should be below 20 kOhm.

Appendix C

The ASCII code

least significant hexdigit of byte	0	1	2	3	4	5	6	7
0	NUL	DLE	SP	0	@	P	`	p
1	SOH	DC1	!	1	A	Q	a	q
2	STX	DC2	"	2	B	R	b	r
3	ETX	DC3	#	3	C	S	c	s
4	EOT	DC4	$	4	D	T	d	t
5	ENQ	NAK	%	5	E	U	e	u
6	ACK	SYN	&	6	F	V	f	v
7	BEL	ETB	'	7	G	W	g	w
8	BS	CAN	(	8	H	X	h	x
9	HT	EM	)	9	I	Y	i	y
A	LF	SUB	*	:	J	Z	j	z
B	VT	ESC	+	;	K	[	k	{
C	FF	FS	,	<	L	\	l	|
D	CR	GS	-	=	M	]	m	}
E	SO	RS	.	>	N	^	n	~
F	SI	US	/	?	O	_	o	DEL

(Column headings: Most significant hexdigit of byte)

Remarks:

1. The most significant bit of the byte is occasionally used as parity check bit. With IPS it serves to indicate an inversion of character and background brightness; e.g. #D3, #D7, #C1, #D0 displays as SWAP

2. The characters of column 0 and 1, and the DEL are non-printing control functions.

3. On teletypes (e.g. ASR 33) the characters of column 6 and 7 are printed by the characters of columns 4 and 5, i.e. upper case only.

Index

(Page numbers as in original PDF, and original pages of PDF under page numbers)
(And on the right here space for own notes)

D

V

W

X

Y

Z

####

Some Links:

AMSAT
https://www.amsat.org/

AMSAT Germany
https://amsat-dl.org/en/

Forth
https://en.wikipedia.org/wiki/Forth_(programming_language)

The Forth Bookshelf - for more information about this language
https://www.amazon.co.uk/Juergen-Pintaske/e/B00N8HVEZM

MPE Forth
www.mpeforth.com

FORTH, Inc.
www.forth.com

Ichigo Jam
(as a possible candidate to implement IPS on a modern ARM Processor, but still a very low-cost and interesting hardware. Either with Video Interface as then – or headless via a USBtoTTL converter and a serial link to a PC as terminal function
https://ichigojam.net/index-en.html

####
IPS Book v5_2019_05_26 Kindle

www.ingramcontent.com/pod-product-compliance
Lightning Source LLC
Chambersburg PA
CBHW030005190526
45157CB00014B/431

* 9 7 8 1 0 9 6 9 9 2 1 5 8 *